COOL CARS

Quentin Willson

HD·WD·34

COOL
CARS

WITHDRAWN FROM STOCK

Quentin Willson

A DORLING KINDERSLEY BOOK

London, New York, Munich, Melbourne, Delhi

THIS EDITION

EDITOR: ALEXANDRA BEEDEN
PROJECT ART EDITOR: LAURA ROBERTS
SENIOR ART EDITOR: HELEN SPENCER
MANAGING ART EDITOR: KAREN SELF
MANAGING EDITOR: ESTHER RIPLEY
PUBLISHER: SARAH LARTER
ART DIRECTOR: PHIL ORMEROD
ASSOCIATE PUBLISHING DIRECTOR:
LIZ WHEELER
PUBLISHING DIRECTOR: JONATHAN METCALF
PRE-PRODUCTION PRODUCER:
REBECCA FALLOWFIELD
SENIOR PRODUCER: GEMMA SHARPE
JACKET DESIGNER: NATALIE GODWIN
JACKET EDITOR: MANISHA MAJITHA
JACKET DESIGN DEVELOPMENT MANAGER:
SOPHIA TAMPAKOPOULOS

DK INDIA

SENIOR ART EDITORS: ANJANA NAIR,
RANJITA BHATTACHARJI
ART EDITOR: DEVAN DAS
ASSISTANT ART EDITORS: ANKITA MUKHERJEE,
NIYATI GOSAIN, PAYAL ROSALIND MALIK
MANAGING ART EDITOR: ARUNESH TALAPATRA
SENIOR EDITOR: SRESHTHA BHATTACHARYA
EDITOR: VIBHA MALHOTRA
MANAGING EDITOR: PAKSHALIKA JAYAPRAKASH
DTP DESIGNERS: VISHAL BHATIA,
SACHIN GUPTA, NEERAJ BHATIA
PRE-PRODUCTION MANAGER:
BALWANT SINGH
PRODUCTION MANAGER: PANKAJ SHARMA

PREVIOUS EDITION

PRODUCED FOR DORLING KINDERSLEY BY
PHIL HUNT (EDITORIAL), MARK JOHNSON DAVIES (DESIGN)

SENIOR EDITOR: EDWARD BUNTING
SENIOR ART EDITOR: KEVIN RYAN
MANAGING EDITOR: SHARON LUCAS
SENIOR MANAGING ART EDITOR: DEREK COOMBES
DTP DESIGNER: SONIA CHARBONNIER
PRODUCTION CONTROLLER: BETHAN BLASE

FIRST PUBLISHED IN GREAT BRITAIN IN 2001
THIS REVISED EDITION PUBLISHED IN 2014 BY DORLING KINDERSLEY LIMITED,
80 STRAND, LONDON WC2R 0RL
A PENGUIN RANDOM HOUSE COMPANY

2 4 6 8 10 9 7 5 3 1
001 - 192885 - Apr/14

COPYRIGHT © 2001, 2014 DORLING KINDERSLEY LIMITED, LONDON
TEXT COPYRIGHT © 2001, 2014 QUENTIN WILLSON

THE RIGHT OF QUENTIN WILLSON TO BE IDENTIFIED AS WRITER OF THIS WORK HAS BEEN
ASSERTED BY HIM IN ACCORDANCE WITH THE COPYRIGHT, DESIGNS, AND PATENTS ACT 1988.

A CIP CATALOGUE RECORD FOR THIS BOOK IS AVAILABLE FROM THE BRITISH LIBRARY
ISBN 978-1-4093-3984-7

Colour reproduced by Colourscan, Singapore
Printed and bound in China by South China

Discover more at
www.dk.com

Note on Specification Boxes: Unless otherwise indicated, all figures pertain to the
particular model in the specification box. Engine capacity for American cars is measured in
cubic inches (cid). A.F.C. is an abbreviation for average fuel consumption.

CONTENTS

INTRODUCTION

Some cars are cool and some have all the sexiness and desirability of an old shoe. Automotive history is littered with dismal failures that were ugly, slow, badly made, drove like donkey carts, or were just plain awful. Some of us are old enough to remember Morris Marinas, Austin Allegros, Ford Pintos, Trabants, Triumph TR7s, Toyota Cedrics, Volkswagen K70s, and Yugos. That such mediocre motors actually made it into production will always be a mystery, but the public weren't fooled and proved it by buying these clunkers in tiny numbers. Cars like these will always stand as monuments to how the automotive industry occasionally gets things dramatically wrong. But mercifully, sometimes, they get it right and produce cars that become hugely desirable icons of cool. And that's what this little book is all about.

Between these pages you'll find a colourful selection of some of the world's greatest cars. Some are breathlessly fast, some are pinnacles of clever design, and some set new technological standards. But all possess that magical allure that makes us want to own and drive them. As I write, the world's appetite for cool cars has never been greater. The market for curvy classics is red hot and hardly a month goes past without another auction price record being broken for an elderly Ferrari or an Aston Martin. Old cars that a couple of decades ago were changing hands for tiny amounts of money have now increased in value as much as a thousand times and become a better investment than gold – literally.

When I first wrote this book I said that many classics were so cheap it seemed criminal. I was amazed that the selling prices of E-Type Jaguars *(see pages 306–09)* and Aston Martin DB4 *(see pages 32–35)* were so ridiculously low. Well, time has proved me right and a lot of the classics I recommended as bargain buys in the first edition of *Cool Cars* have since mushroomed in price to insanely stratospheric levels. I hate to say "I told you so", but if you'd done as I suggested and bought an Aston Martin DB4 and Ferrari Daytona *(see page 233)* for around £70,000 in 2001, the pair would now be worth over a million today. In a little over a decade the desire to own distinctive and rare classics has become an unstoppable market force worth many billions.

But our obsession with cool cars doesn't stop at the old stuff. The market for high-tab, glamorous new cars is red hot too. And the market is cooking all over the world. Modern Bentleys, Rolls Royces, Ferraris, Aston Martins, Maseratis, and McLarens have almost become celebrities in their own right. There are waiting lists for Jaguars and Range Rovers, over-list premiums being paid for Ferraris, and queues of desperate buyers chasing used luxury and sports cars. And given that the world is suffering under the worst recession since the Second World War, this simply should not be happening. Our desire for distinctive sexy wheels is probably the most powerful it's ever been in the history of the motor car and many of us are willing to spend all the money we haven't got just to get behind the wheel of a cool ride.

What is even more surprising is that a blizzard of anti-car legislations, dire warnings about climate change, growing congestion, and pressure from environmentalists hasn't really changed our attitudes. Most people with a soul, when given the choice between a Toyota Prius hybrid and a BMW 3 Series, will always go for the BMW. The Prius may be an enormously clever car but it just can't tickle our hearts with the same delicate fingers as the BMW. And that's because a cool car defines who we are in the social pecking order. No other symbol in society changes other people's perceptions of us like cars – they're mobile, they're visible, and they're a currency that almost everybody understands. And if your ride is cool, onlookers just sigh in admiration. Being bland and predictable won't get you those coveted looks of warm approval and it doesn't matter if your cool car is ancient or modern, you'll be making that very important statement that you chose to be different from everybody else.

So let's sit back and relish this new golden age of motoring where there are so many wonderful cars to choose from. In all the years I've been writing and broadcasting about motoring, I can't remember a time when the choice of desirable and genuinely charismatic cars was so amazingly huge. The selection of cars in this book may be eclectic but I guarantee that all of them, without exception, will turn heads and disarm and charm in equal measure.

And remember this: our love of cool cars isn't going to go away anytime soon so if you've got the cash, think about buying yourself a cool classic. There are very few things that you can buy in life that can offer the same level of enjoyment and fun as an old car that steadily increases in value over the years. It may be too late to buy that cheap Daytona or DB4 but there are plenty of other classics (and modern neo-classics) that are still affordable, still sexy, and still special. Go on, change your life and buy a cool car. I promise that you won't regret it and you'll be getting the keys to a new world of like-minded enthusiasts, all of whom refuse to drive something dreary, plain, or beige. Enjoy the ride.

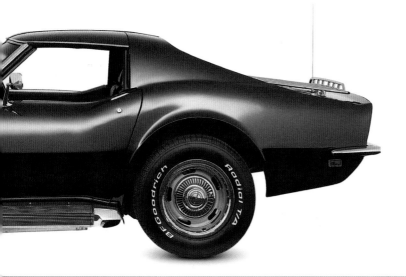

AC *Ace-Bristol*

AGONIZINGLY PRETTY, THE AC ACE catapulted the homespun Thames Ditton company into the motoring limelight, instantly earning it a reputation as makers of svelte sports cars for the tweedy English middle classes. Timelessly elegant, swift, poised, and mechanically uncomplicated, the Ace went on to form the platform for the legendary AC Cobra *(see pages 16–19)*. Clothed in a light alloy body and powered by a choice of AC's own delicate UMB 2.0 unit, the hardier 2.0 Bristol 100D2 engine, or the lusty 2.6 Ford Zephyr power plant, the Ace drove as well as it looked. Its shape has guaranteed the Ace a place in motoring annals. Chaste, uncluttered, and simple, it makes a Ferrari look top-heavy and clumsy. Purists argue that the Bristol-powered version is the real thoroughbred Ace, closest to its original inspiration, the Bristol-powered Tojeiro prototype of 1953.

SIDE VIEW

The most handsome British roadster of its day, and as lovely as an Alfa Romeo Giulietta Sprint, the Ace had an Italianate simplicity. Proof of the dictum that less is more, the Ace's gently sweeping profile is a triumph of form over function.

SIDESCREEN

Folding perspex sidescreens helped to prevent turbulence in the cockpit at high speed. ⎯⎯⎯⎯⎯

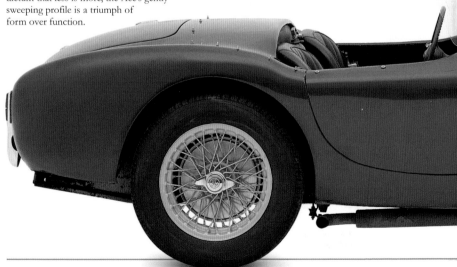

IMPRESSIVE SPEC
The Ace had triple Solex carbs, push-rod overhead valve gear, a light alloy head, and a cast-iron crankcase.

BRASS PLATE
The firing order of the Ace's six cylinders was displayed on an engine plate.

FIRING ORDER
1 5 3 6 2 4
Bristol

BONNET CATCHES
Forward-hinged bonnet was locked by two chrome catches, opened by a small T-shaped key.

ENGINE
Shared by the BMW 328, the hemi-head 125 bhp 2-litre Bristol engine was offered as a performance conversion for the Ace.

SPECIFICATIONS

MODEL AC Ace-Bristol (1956–61)

PRODUCTION 463

BODY STYLE Two-door, two-seater sports roadster.

CONSTRUCTION Space-frame chassis, light alloy body.

ENGINE Six-cylinder push-rod 1971cc.

POWER OUTPUT 105 bhp at 5000 rpm (optional high-performance tune 125 bhp at 5750 rpm).

TRANSMISSION Four-speed manual Bristol gearbox (optional overdrive).

SUSPENSION Independent front and rear with transverse leaf spring and lower wishbones.

BRAKES Front and rear drums. Front discs from 1957.

MAXIMUM SPEED 188 km/h (117 mph)

0–60 MPH (0–96 KM/H) 9.1 sec

0–100 MPH (0–161 KM/H) 27.2 sec

A.F.C. 7.6 km/l (21.6 mpg)

BRAKES
Front discs were an option in 1957, but later standardized.

PROPORTIONS

The AC was simplicity itself – a box for the engine, a box for the people, and a box for the luggage. On the handling side, production cars used Bishop cam-and-gear steering, which gave a turning circle of 11 m (36 ft), and required just two deft turns of the steering wheel lock-to-lock.

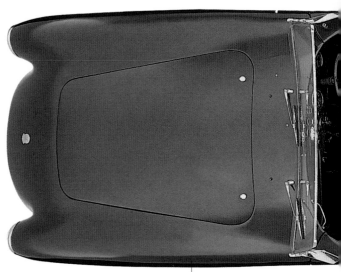

SHARED WHEEL
Steering wheel was shared with the Austin Healey (see pages 48–55) and the Daimler SP Dart (see pages 190–93).

CONSTRUCTION
Known as Superleggera *construction, a network of steel tubes was covered by aluminium panels, based on the outline of the 1949 Ferrari 122.*

EXPORT SUCCESS
The Ace became one of AC's most successful creations, with a huge proportion exported to America, where its character as an English cad's crumpet-catcher justified its price tag of a small house.

COOLING
The Ace's wide, toothy grin fed air into the large radiator that was shared by the AC two-litre saloon.

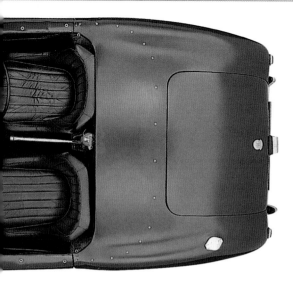

INTERIOR

In pure British tradition, the Ace's cockpit was stark, with gauges and switches haphazardly pebble-dashed across the fascia. The two larger dials were a speedometer – with a clock inset into the dial – and a rev-counter.

TONNEAU FASTENERS

For die-hards who always drove with the hood down, a tonneau cover could be fitted which kept your feet warm while your face froze.

REAR-ENGINED GUSTO

Engines were placed well back and gave an 18 per cent rearward bias to the weight distribution. Performance-wise, it helped – an Ace recorded an average of 156 km/h (97 mph) over 3,781 km (2,350 miles) at the 1957 Le Mans 24 Hours, the fastest ever for a Bristol-engined car.

REVISED LIGHTS

Later Aces had a revised rear deck, with square tail lights and a bigger boot.

AC *Cobra 427*

AN UNLIKELY ALLIANCE BETWEEN AC CARS, a traditional British car maker, and Carroll Shelby, a charismatic Texan racer, produced the legendary AC Cobra. AC's sports car, the Ace *(see pages 12–15)* was turned into the Cobra by shoe-horning in a series of American Ford V8s, starting with 4.2 and 4.7 Mustang engines. In 1965, Shelby, always a man to take things to the limit, squeezed in a thunderous 7-litre Ford engine, in an attempt to realize his dream of winning Le Mans. Although the 427 was not fast enough to win and failed to sell in any quantity, it was soon known as one of the most aggressive and romantic cars ever built. GTM 777F at one time held the record as the world's fastest accelerating production car and in 1967 was driven by the British journalist John Bolster to record such Olympian figures as an all-out maximum of 265 km/h (165 mph) and a 0–60 (96 km/h) time of an unbelievable 4.2 seconds.

MUSCLY PROFILE
The 427 looked fast standing still. Gone was the lithe beauty of the original Ace, replaced by bulbous front and rear arches, fat 19-cm (7½-in) wheels, and rubber wide enough to roll a cricket pitch.

WHEELS
Initially pin-drive Halibrand magnesium alloy but changed for Starburst wheels (designed by Shelby employee Pete Brock) when supplies dried up.

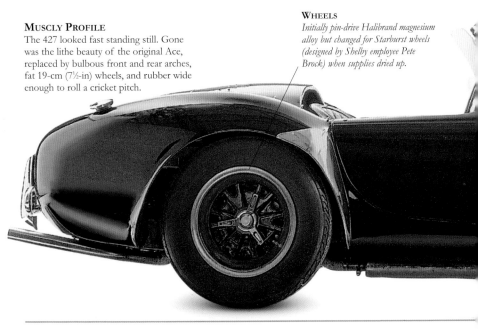

BODYWORK
The Cobra's body was constructed from hand-rolled aluminium wrapped around a tubular steel frame, which proved very light yet extremely strong.

BUMPERS
Bumpers were token chromed tubes, with the emphasis on saving weight.

EXHAUST
Racing Cobras usually had side exhausts, which increased power and noise.

SIDESCREENS
Small perspex sidescreens helped cut down wind noise.

COOLING
Wing vents helped reduce brake and engine temperatures.

EXTRA HORSEPOWER
Competition and semi-competition versions with tweaked engines could exceed 500 bhp.

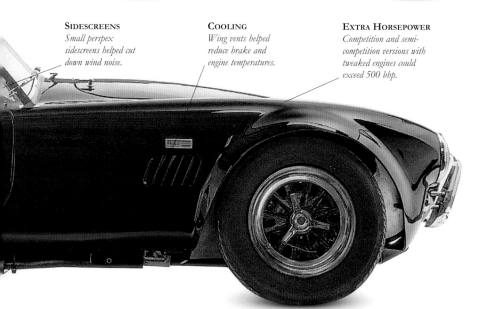

ENGINE
The mighty 7-litre 427 block had years of NASCAR (National Association of Stock Car Automobile Racing) racing success and easily punched out power for hours. The street version output ranged from 300 to 425 bhp.

FRAME
The windscreen frame was hand-made and polished.

RADIATOR TANK
Radiator header tank kept things cool, helped by twin electric fans.

AIR CLEANER
Under the massive air cleaner are two large four-barrel carburettors.

TYRES
Cobra tyres were always Goodyear as Shelby was a long-time dealer.

UPGRADED CHASSIS
The chassis was virtually all new and three times stronger than the earlier Cobra 289's, with computer-designed anti-dive and anti-squat characteristics. Amazingly, the 289's original Salisbury differential proved more than capable of handling the 427's massive wall of torque.

POCKET DYNAMO
Even the "baby" 4.7
Cobras – as seen in this
contemporary poster –
were good for 222 km/h
(138 mph) and could squeal
up to 60 mph (96 km/h) in
under six seconds.

ENGINE CHANGES
*Early Cobras had
260cid engines. Later
cars were fitted with
Mustang 289 V8s.*

SPECIFICATIONS

MODEL AC Cobra 427 (1965–68)

PRODUCTION 316

BODY STYLE Light alloy, two-door, two-seater, open sports.

CONSTRUCTION Separate tubular steel chassis with aluminium panels.

ENGINE V8, 6989cc.

POWER OUTPUT 425 bhp at 6000 rpm.

TRANSMISSION Four-speed all-synchromesh.

SUSPENSION All-round independent with coil springs.

BRAKES Four-wheel disc.

MAXIMUM SPEED 265 km/h (165 mph)

0–60 MPH (0–96 KM/H) 4.2 sec

0–100 MPH (0–161 KM/H) 10.3 sec

A.F.C. 5.3 km/l (15 mpg)

INTERIOR
The interior was basic, with traditional
1960s British sports car features of
black-on-white gauges, small bucket
seats, and wood-rim steering wheel.

AC 428

THE AC 428 NEEDS A NEW word of its very own – "brutiful" perhaps, for while its brute strength derives from its Cobra forebear, the 428 has a sculpted, stately beauty. This refined bruiser was born of a thoroughbred cross-breed of British engineering, American power, and Italian design. The convertible 428 was first seen at the London Motor Show in October 1965; the first fixed-head car – the so-called fastback – was ready in time for the Geneva Motor Show in March 1966. But production was beset by problems from the start; first cars were not offered for sale until 1967, and as late as March 1969, only 50 had been built. Part of the problem was that the 428 was priced between the more expensive Italian Ferraris and Maseratis and the cheaper British Astons and Jensens. Small-scale production continued into the 1970s, but its days were numbered and it was finally done for by the fuel crisis of October 1973; the last 428 – the 80th – was sold in 1974.

ITALIAN STYLING
Styled by Pietro Frua in Turin, the AC 428 was available in both convertible and fixed-head fastback form. It was based on an AC Cobra 427 chassis, virtually standard apart from a 15-cm (6-in) increase in wheelbase.

THIN SKINNED
Early cars had aluminium doors and bonnet; later models were all-steel.

CONTEMPORARY LOOKS
The design contains subtle reminders of a number
of contemporary cars, not least the Maserati Mistral.
Hardly surprising really, since the Mistral was
also penned by Pietro Frua.

SPECIFICATIONS

MODEL AC 428 (1966–73)

PRODUCTION 80 (51 convertibles,
29 fastbacks)

BODY STYLES Two-seat convertible
or two-seat fastback coupé.

CONSTRUCTION Tubular-steel backbone
chassis/separate all-steel body.

ENGINES Ford V8, 6997cc or 7016cc.

POWER OUTPUT 345 bhp at 4600 rpm.

TRANSMISSION Ford four-speed manual
or three-speed auto; Salisbury rear axle
with limited-slip differential.

SUSPENSION Double wishbones and
combined coil spring/telescopic damper
units front and rear.

BRAKES Servo-assisted Girling discs front
and rear.

MAXIMUM SPEED 224 km/h
(139.3 mph) (auto)

0–60 MPH (0–96 KM/H) 5.9 sec (auto)

0–100 MPH (0–161 KM/H) 14.5 sec

A.F.C. 4.2–5.3 km/l (12–15 mpg)

AIR VENTS
*In an effort to combat engine
overheating, later cars have
air vents behind the wheels.*

ALL LACED UP
*Standard wheels were substantial triple-laced,
wired-up affairs, secured by a three-eared nut.*

WEATHER BEATER

The hood was tucked under a cover which, in early models, was made of metal. When up, the hood made the cockpit feel somewhat claustrophobic, but the plastic rear "window" was generously proportioned.

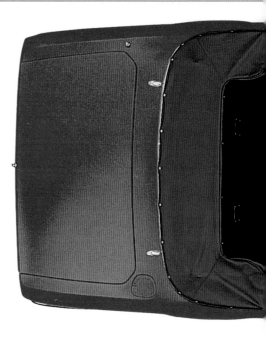

DASHBOARD

Switchgear may be scattered around like confetti, but the instruments are grouped clearly in front of the driver. The speedo *(far left)* reads to an optimistic 180 mph (290 km/h), while the tachometer *(far right)* limits at 8000 rpm.

REAR VIEW

The 428 may have been a refined muscle car, but it was not totally unique; it featured parts from other manufacturers, such as rear lights from Fiat.

DESIGN CREDIT
Frua are credited with a "Creazione Frua" badge on each wing.

POWER UNIT
Pre-1967, the car used the same 427 cubic inch (6998cc) V8 as the Cobra *(see pages 16–19)*, so was originally known as the AC 427.

ALFA ROMEO *1300 Junior Spider*

DRIVEN BY DUSTIN HOFFMAN TO THE strains of Simon and Garfunkel in the film *The Graduate*, the Alfa Spider has become one of the most accessible cult Italian cars. This is hardly surprising when you consider the little Alfa's considerable virtues: a wonderfully responsive all-alloy, twin-cam engine, accurate steering, sensitive brakes, a finely balanced chassis, plus matinee idol looks. Not for nothing has it been called the poor man's Ferrari. First launched at the Geneva Motor Show in 1966, Alfa held a worldwide competition to find a name for their new baby. After considering 140,000 entries, with suggestions like Lollobrigida, Bardot, Nuvolari, and even Stalin, they settled on Duetto, which neatly summed up the car's two's-company-three's-a-crowd image. Despite the same price tag as the much faster and more glamorous Jaguar E-Type *(see pages 306–09)*, the Spider sold over 100,000 units during its remarkable 26-year production run.

CONTEMPORARY CONTROVERSY
One of Pininfarina's last designs, the Spider's rounded front and rear and deep-channelled scallop running along the sides attracted plenty of criticism. One British motoring magazine dubbed it "compact and rather ugly".

HOOD
The Spider's hood was beautifully effective. It could be raised with only one arm without leaving the driver's seat.

LOGO
Pininfarina's credit was indicated by his logo.

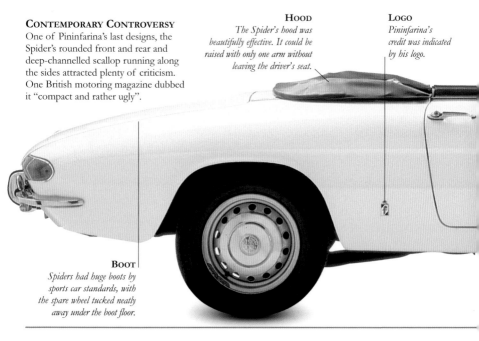

BOOT
Spiders had huge boots by sports car standards, with the spare wheel tucked neatly away under the boot floor.

INTERIOR

The dashboard was painted metal up to 1970. Minor controls were on fingertip stalks, while the windscreen wipers had an ingenious foot button positioned on the floor.

SPECIFICATIONS

MODEL Alfa Romeo 1300 Junior Spider (1968–78)

PRODUCTION 7,237

BODY STYLE Two-door, two-seater.

CONSTRUCTION All-steel monocoque body.

ENGINE All-alloy twin-cam 1290cc.

POWER OUTPUT 89 bhp at 6000 rpm.

TRANSMISSION Five-speed.

SUSPENSION *Front:* independent; *Rear:* live axle with coil springs.

BRAKES Four-wheel disc.

MAXIMUM SPEED 170 km/h (106 mph)

0–60 MPH (0–96 KM/H) 11.2 sec

0–100 MPH (0–161 KM/H) 21.3 sec

A.F.C. 10.3 km/l (29 mpg)

RACING ALFA
The later Alfa Romeo Montreal had a race-bred 2.5 V8 that gave a top speed of 225 km/h (140 mph).

HEADLAMPS
Perspex headlamp covers were banned in the US and were never fitted to the 1300 Juniors.

WHEEL ARCHES
The lack of rustproofing meant that the arches were prone to decay.

DRIVING POSITION
*All Spider cockpits had
the Italian ape-like
driving position – long
arms and short legs.*

QUALITY TAIL
The "boat-tail" rear was shared by all Spiders up to
1970 and is the styling favoured by Alfa purists. It
was replaced by a squared-off Kamm tail.

BODYWORK
*The Spider's bodywork
corroded alarmingly
quickly due to the poor-
quality steel.*

STYLISH AND COOL
The Spider has to be one of Alfa's great
post-war cars, not least because of its
contemporary design. It was penned
by Battista Pininfarina, the founder
of the renowned Turin-based
design house.

NOSE SECTION
*Disappearing nose
was very vulnerable
to parking dents.*

WMT 97G

ALFA'S BAMBINO
The 1300 Junior was the baby of
the Spider family, introduced in
1968 to take advantage of Italian
tax laws. As well as the "Duetto",
which refers to 1600 Spiders,
there was also a 1750cc model
in the range. Large production
numbers and high maintenance
costs mean that prices of Spiders
are invitingly low.

STYLISH GRILLE
This hides a twin-cam,
energy-efficient engine with
hemispherical combustion
chambers. Some of the
mid-'70s Spiders imported to
the US, however, were overly
restricted; the catalysed 1750,
for example, could only manage
a miserly top speed of just
159 km/h (99 mph).

AMC *Pacer*

THE 1973 FUEL CRISIS HIT America's psyche harder than the Russians beating them to space in the Fifties. Cheap and unrestricted personal transport had been a way of life, and then suddenly America faced the horrifying prospect of paying more than forty cents a gallon. Overnight, shares in motor manufacturers became as popular as Richard Nixon. Detroit's first response was to kill the muscle car dead. The second was to revive the "compact" and invent the "sub-compact". AMC had first entered the sub-compact market in 1970 with its immensely popular Gremlin model, but the 1975 Pacer was a different beast. Advertised as "the first wide small car", it had the passenger compartment of a sedan, the nose of a European commuter shuttle, and no back end at all. Ironically, it wasn't even that economical, but America didn't notice because she was on a guilt trip, buying over 70,000 of the things in '75 alone.

WINDSCREEN
The aerodynamic windscreen aided fuel economy and reduced interior noise.

STYLING TO TALK ABOUT
In the mid-Seventies, the Pacer was sold as the last word; "the face of the car of the 21st century" bragged the ads. Happily, they were wrong. Pundits of the time called it a "football on wheels" and a "big frog".

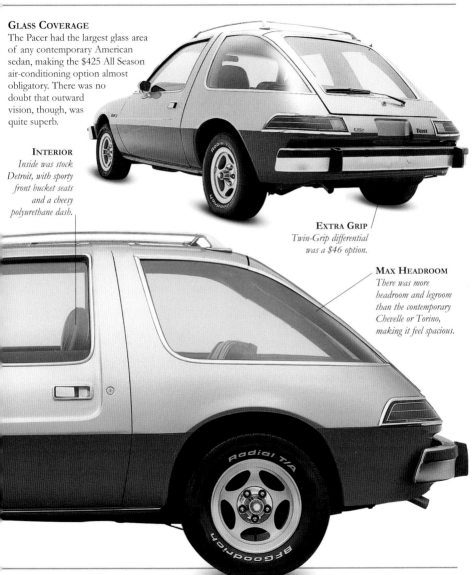

GLASS COVERAGE
The Pacer had the largest glass area of any contemporary American sedan, making the $425 All Season air-conditioning option almost obligatory. There was no doubt that outward vision, though, was quite superb.

INTERIOR
Inside was stock Detroit, with sporty front bucket seats and a cheesy polyurethane dash.

EXTRA GRIP
Twin-Grip differential was a $46 option.

MAX HEADROOM
There was more headroom and legroom than the contemporary Chevelle or Torino, making it feel spacious.

EXTRA WIDTH

The body was almost as wide as it was long, and though opinion was divided on the Pacer's looks, it did garner some hefty praise; *Motor Trend* magazine called the styling "the most innovative of all US small cars". Credit went to Richard Teague, who also penned the '84 Jeep Cherokee.

BOOT SPACE
With rear seat folded, cargo area was an impressive 30 cubic feet.

FRONT RECLINERS
Adaptability even stretched to the front of the car; 26 per cent of all Pacers had reclining front seats.

LATER LENGTH
In 1977 Pacers were stretched a further 10 cm (4 in) and offered as station wagons.

COSTLY EXTRAS

Surprisingly, the Pacer was never a cheap car. Add a few interior options and air-conditioning and you could easily have been presenting the dealer with a cheque for $5,000. De Luxe trim pack included wood effect side and rear panels, which made the Pacer about as tasteful as Liberace.

STEERING
The Pacer's rack-and-pinion steering was one of the first on a US car.

SPECIFICATIONS

MODEL AMC Pacer

PRODUCTION 72,158 (1975)

BODY STYLE Three-door saloon.

CONSTRUCTION Steel unitary body.

ENGINES 232cid, 258cid sixes.

POWER OUTPUT 90–95 bhp.

TRANSMISSION Three-speed manual with optional overdrive, optional three-speed Torque-Command automatic.

SUSPENSION *Front:* coil springs; *Rear:* semi-elliptic leaf springs.

BRAKES Front discs, rear drums.

MAXIMUM SPEED 169 km/h (105 mph)

0–60 MPH (0–96 KM/H) 14 sec

A.F.C. 6.4–8.5 km/l (18–24 mpg)

REAR INSPIRATION
Unbelievably, the Pacer's rear end inspired the comely rump of the Porsche 928.

PACER POWER
Stock power was a none-too-thrifty 258cid straight-six unit. In addition, for all its eco pretensions, you could still specify a 304cid V8.

BUMPERS
Originally slated to use urethane bumpers, production Pacers were fitted with steel versions to save money.

ASTON MARTIN *DB4*

THE DEBUT OF THE DB4 IN 1958 heralded the beginning of the Aston Martin glory years, ushering in the breed of classic six-cylinder DB Astons that propelled Aston Martin on to the world stage. Earlier post-war Astons were fine sporting enthusiasts' road cars, but with the DB4 Astons acquired a new grace, sophistication, and refinement that was, for many, the ultimate flowering of the grand tourer theme. The DB4 looked superb and went like stink. The DB5, which followed, will forever be remembered as the James Bond Aston, and the final expression of the theme came with the bigger DB6. The cars were glorious, but the company was in trouble. David Brown, the millionaire industrialist owner of Aston Martin and the DB of the model name, had a dream. But, in the early Seventies, with losses of £1 million a year, he bailed out of the company, leaving a legacy of machines that are still talked about with reverence as the David Brown Astons.

DASHBOARD
The fascia is a gloriously unergonomic triumph of form over function; gauges are scattered all over an instrument panel deliberately similar to the car's grinning radiator grille.

IN THE MIRROR
Dipping rear-view mirror was also found in many Jaguars of the period.

BRITISH LIGHTWEIGHT
Superleggera, Italian for "super-lightweight", refers to the technique of body construction: aluminium panels rolled over a framework of steel tubes.

UNHINGED
First-generation DB4s had a rear-hinged bonnet.

SPECIFICATIONS

MODEL Aston Martin DB4 (1958–63)

PRODUCTION 1,040 (fixed head); 70 (convertible); 95 (fixed-head DB4 GTs).

BODY STYLES Fixed-head coupé or convertible.

CONSTRUCTION Pressed-steel and tubular inner chassis frame, with aluminium-alloy outer panels.

ENGINES Inline six 3670cc/3749cc.

POWER OUTPUT 240 bhp at 5500 rpm.

TRANSMISSION Four-speed manual (with optional overdrive).

SUSPENSION *Front:* independent by wishbones, coil springs and telescopic dampers; *Rear:* live axle located by trailing arms and Watt linkage with coil springs and lever-arm dampers.

BRAKES Four-wheel disc.

MAXIMUM SPEED 225+ km/h (140+ mph)

0–60 MPH (0–96 KM/H) 8 sec

0–100 MPH (0–161 KM/H) 20.1 sec

A.F.C. 3.6–7.8 km/l (14–22 mpg)

ASTON SMILE
The vertical bars in this car's radiator grille show it to be a so-called Series 4 DB4, built between September 1961 and October 1962.

NO PRETENSIONS
There is no doubt that the DB4 has got serious attitude. Its lines may be Italian, but it has none of the dainty delicacy of some contemporary Ferraris and Maseratis – the Aston's spirit is somehow true-Brit.

BOOT PANEL
Complex curves meant the boot lid was one of the most difficult-to-produce panels in the entire car. Their hand-built nature means no two Astons are alike.

LIGHTS
Rear lights and front indicators were straight off the utilitarian Land Rover.

BOND MATERIAL
The DB4's stance is solid and butch, but not brutish – more British Boxer than lumbering Bulldog, aggressive yet refined. It is an ideal blueprint for a James Bond car.

CLASSIC STYLING
Clothed in an Italian body by Carrozzeria Touring of Milan, the DB4 possessed a graceful yet powerful elegance. Under the aluminium shell was Tadek Marek's twin-cam straight-six engine, which evolved from Aston's racing programme.

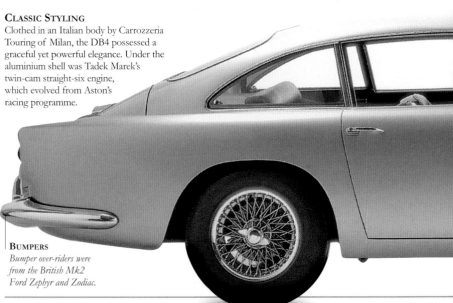

BUMPERS
Bumper over-riders were from the British Mk2 Ford Zephyr and Zodiac.

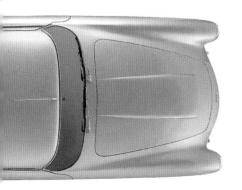

ENGINE

It looks very much like the contemporary Jaguar XK twin-cam straight-six, but Tadek Marek's design is both more powerful and vastly more complicated. Triple SU carburettors show this to be a Vantage engine with larger valves and an extra 20 bhp.

UPHOLSTERY

While rear seats in the fixed-head offer limited space, just look at the richness and quality of the Connolly leather. The ride wasn't quite as impressive, though – rear suspension was through basic lever-arm units.

SUSPENSION

Front suspension was double wishbones with coil springs and telescopic dampers.

ASTON MARTIN *V8*

A NEAR TWO-TONNE GOLIATH powered by an outrageous hand-made 5.3-litre engine, the DBS V8 was meant to be Aston's money-earner for the 1970s. Based on the six-cylinder DBS of 1967, the V8 did not appear until April 1970. With a thundering 257 km/h (160 mph) top speed and incredible sub seven-second 0–60 time, Aston's new bulldog instantly earned a place on every millionaire's shopping list. The trouble was that it drove into a worldwide recession – in 1975 the Newport Pagnell factory produced just 19 cars. Aston's bank managers were worried men, but the company pulled through. The DBS became the Aston Martin V8 in 1972 and continued until 1989, giving birth to the legendary 400 bhp Vantage and gorgeous Volante Convertible. Excessive, expensive, impractical, and impossibly thirsty, the DBS V8 and AM V8 are wonderful relics from a time when environmentalism was just another word in the dictionary.

NEW CONSTRUCTION

DBS was one of the first Astons with a chassis and departed from the traditional Superleggera tubular superstructure of the DB4, 5, and 6. Like Ferraris and Maseratis, Aston prices were ballyhooed up to stratospheric levels in the Eighties.

ASTON LINES
Smooth tapering cockpit line is an Aston hallmark echoed in the current DB7.

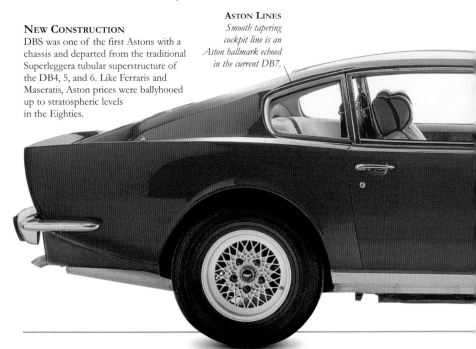

REAR ASPECT
Prodigious rear overhang makes the rear aspect look cluttered.

REAR WINDOW
Thin rear window gave the driver limited rearward vision.

SPOILER
Discreet rear spoiler was part of the gently sweeping wing line.

TWIN PIPES
Hand-made bumpers covered huge twin exhausts – a gentle reminder of this Aston's epic V8 grunt.

B391 AJD

OWNERS WITH PEDIGREE
Cars with incredible presence, Astons were good enough for James Bond, King Hussein of Jordan, Peter Sellers, and even the Prince of Wales – who has owned a DB6 Volante from new.

POWER BULGE
Massive bonnet power bulge was to clear four carburettors.

ENGINE

The alloy V8 was first seen in Lola sports-racing cars. The massive air-cleaner box covers a quartet of twin-choke Weber carbs, which guzzle one litre of fuel for every 4.6 km (13 mpg), and much less if you enjoy yourself.

EIGHTIES' PRICE

In the Eighties, top quality DBSs changed hands for £50,000 plus

POWER UNIT

V8's engine churned out over 300 bhp, but later models could boast 400 bhp.

FRONT END

Shapely "cliff-hanger" nose was always a DBS trademark.

BOND CAR

A 1984 AM V8 Volante featured in the James Bond film *The Living Daylights*, with Timothy Dalton. In 1964 a DB5 was the first Aston to star alongside James Bond in the film *Goldfinger*, this time with Sean Connery.

SPOILER

Chin spoiler and undertray helped reduce front-end lift at speed.

B391 A

CLASSY CABIN

Over the years the DBS was skilfully updated, without losing its traditional ambience. Features included hide and timber surroundings, air-conditioning, electric windows, and radio cassette. Nearly all V8s were ordered with Chrysler TorqueFlite auto transmission.

BODYWORK
V8's aluminium body was hand-smoothed and lovingly finished.

SUMPTUOUS FITTINGS
As with most Astons, the interior was decked out in the finest quality hide and timber.

SPECIFICATIONS

MODEL Aston Martin V8 (1972–89)

PRODUCTION 2,842 (including Volante and Vantage)

BODY STYLE Four-seater coupé.

CONSTRUCTION Aluminium body, steel platform chassis.

ENGINE Twin OHC alloy 5340cc V8.

POWER OUTPUT Never released but approx 345 bhp (Vantage 400 bhp).

TRANSMISSION Three-speed auto or five-speed manual.

SUSPENSION Independent front, De Dion rear.

BRAKES Four-wheel disc.

MAXIMUM SPEED 259 km/h (161 mph); 278 km/h (173 mph) (Vantage)

0–60 MPH (0–96 KM/H) 6.2 sec (Vantage 5.4 sec)

0–100 MPH (0–161 KM/H) 14.2 sec (Vantage 13 sec)

A.F.C. 4.6 km/l (13 mpg)

AUDI *Quattro Sport*

ONE OF THE RAREST and most iconic Audis ever built was the 250 km/h (155 mph) Quattro Sport. With a short wheelbase, all-alloy 300 bhp engine, and a body made of aluminium-reinforced glass-fibre and Kevlar, it has all the charisma, and nearly all the performance, of a Ferrari GTO. The Quattro changed the way we think about four-wheel drive. Before 1980, four-wheel drive systems had foundered through high cost, weight, and lousy road behaviour. Everybody thought that if you bolted a four-wheel drive system onto a performance coupé it would have ugly handling, transmission whine, and an insatiable appetite for fuel. Audi's engineers proved that the accepted wisdom was wrong and, by 1982, the Quattro was a World Rally Champion. Gone but not forgotten, the Quattro Sport is now a much admired collectors' item.

FUNCTIONAL INTERIOR
While the dashboard layout is nothing special, everything is typically Germanic – clear, tidy, and easy to use. The only touch of luxury in the Quattro is half-leather trim.

RALLY SUCCESS
In competition trim, Audi's remarkable turbocharged engine was pushing out 400 bhp and, by 1987, the fearsome S1 Sport generated 509 bhp. To meet Group B homologation requirements, only 220 Sports were built, all in left-hand drive guise, and only a few destined for sale to some very lucky private owners.

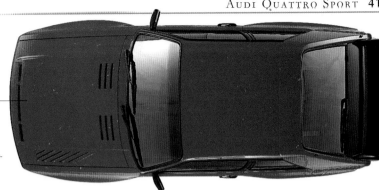

BONNET
Long nose and bonnet bulge cover the intercooler for the turbo unit.

ROOF
Roof sections were made of aluminium-bonded glass-fibre.

HAND-CRAFTED BODY
Bodyshells were welded by a team of just 22 craftsmen.

HOT PROPERTY

From any angle the Quattro Sport is testosterone on wheels, with a butch and aggressive four-square stance. The breeze-block styling, though, meant that the Quattro's aerodynamics were poor.

SPECIFICATIONS

MODEL Audi Quattro Sport (1983–87)

PRODUCTION 220 (all LHD)

BODY STYLE Two-seater, two-door coupé.

CONSTRUCTION Monocoque body from Kevlar, aluminium, glass-fibre, and steel.

ENGINE 2133cc five-cylinder turbocharged.

POWER OUTPUT 304 bhp at 6500 rpm.

TRANSMISSION Five-speed manual, four-wheel drive.

SUSPENSION All-round independent.

BRAKES Four-wheel vented discs with switchable ABS.

MAXIMUM SPEED 250 km/h (155 mph)

0–60 MPH (0–96 KM/H) 4.8 sec

0–100 MPH (0–161 KM/H) 13.9 sec

A.F.C. 6 km/l (17 mpg)

LIMITED EDITION
Of the 1,700 Audis produced each day in the mid-1980s, only three were Quattros, and of a year's output only a tiny amount were Sports.

REAR LIGHTS
Darkened rear lights were included across the whole Quattro range in 1984.

FOUR-SEATER?
While it looked like a four-seater, in practice only two could fit in.

ARCHES
Box wheelarches are a Quattro hallmark, and essential to cover the fat 9Jx15 wheels.

RIDE QUALITY
Though the ride was harder than on normal Quattros, steering was quicker.

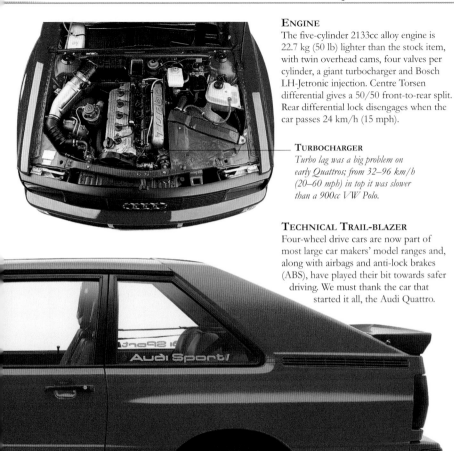

ENGINE

The five-cylinder 2133cc alloy engine is 22.7 kg (50 lb) lighter than the stock item, with twin overhead cams, four valves per cylinder, a giant turbocharger and Bosch LH-Jetronic injection. Centre Torsen differential gives a 50/50 front-to-rear split. Rear differential lock disengages when the car passes 24 km/h (15 mph).

TURBOCHARGER

Turbo lag was a big problem on early Quattros; from 32–96 km/h (20–60 mph) in top it was slower than a 900cc VW Polo.

TECHNICAL TRAIL-BLAZER

Four-wheel drive cars are now part of most large car makers' model ranges and, along with airbags and anti-lock brakes (ABS), have played their bit towards safer driving. We must thank the car that started it all, the Audi Quattro.

AUSTIN *Mini Cooper*

THE MINI COOPER WAS ONE of Britain's great motor sport legends, an inspired confection that became the definitive rally car of the Sixties. Because of its size, manoeuvrability, and front-wheel drive, the Cooper could dance around bigger, more unwieldy cars and scuttle off to victory. The hot Mini was a perfect blend of pin-sharp steering, terrific handling balance, and a feeling that you could get away with almost anything. Originally the brainchild of racing car builder John Cooper, the Mini's designer, Alec Issigonis, thought it should be a "people's car" rather than a performance machine and did not like the idea of a tuned Mini. Fortunately BMC (British Motor Corporation) did, and agreed to a trial run of just 1,000 cars. One of their better decisions.

SPOTLIGHT
Roof-mounted spotlight could be rotated from inside the car.

TYRES
Radial tyres were on the Cooper S but not the standard Cooper.

RALLY REAR

24 PK wears the classic Mini rally uniform of straight-through exhaust, Minilite wheels, roll bar, twin fuel tanks, and lightweight stick-on number plates. BMC had a proactive Competitions Department.

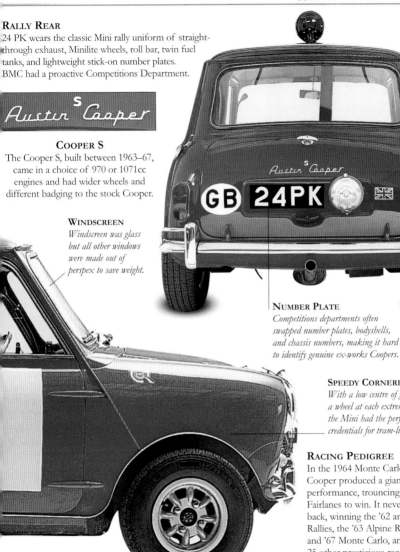

COOPER S

The Cooper S, built between 1963–67, came in a choice of 970 or 1071cc engines and had wider wheels and different badging to the stock Cooper.

WINDSCREEN
Windscreen was glass but all other windows were made out of perspex to save weight.

NUMBER PLATE
Competitions departments often swapped number plates, bodyshells, and chassis numbers, making it hard to identify genuine ex-works Coopers.

SPEEDY CORNERING
With a low centre of gravity and a wheel at each extreme corner, the Mini had the perfect credentials for tram-like handling.

RACING PEDIGREE
In the 1964 Monte Carlo Rally, the Cooper produced a giant-killing performance, trouncing 4.7-litre Fairlanes to win. It never looked back, winning the '62 and '64 Tulip Rallies, the '63 Alpine Rally, the '65 and '67 Monte Carlo, and more than 25 other prestigious races.

RARE BLOCK
Though this is a 1071cc example, the 970cc version was the rarest of all Coopers, with only 964 made.

ENGINE
The 1071cc A-series engine would rev to 7200 rpm, producing 72 bhp. Crankshaft, con-rods, valves, and rockers were all toughened, and the Cooper also had a bigger oil pump and beefed-up gearbox. Lockheed disc brakes and servo provided the stopping power.

GRILLE
Front grille was quick-release to allow access for emergency repairs to distributor, oil cooler, starter motor, and generator.

RACE EXPERIENCE
This example, 24 PK, was driven by Sir Peter Moon and John Davenport in the 1964 Isle of Man Manx Trophy Rally. But, while leading the pack on the penultimate stage of the rally at Druidale, 24 PK was badly rolled and needed a complete reshell. Many works Coopers led a hard life, often rebuilt and reshelled several times.

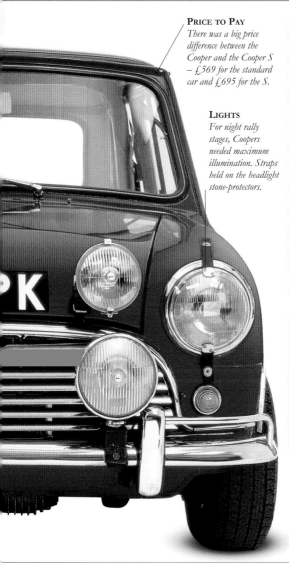

PRICE TO PAY
*There was a big price
difference between the
Cooper and the Cooper S
– £569 for the standard
car and £695 for the S.*

LIGHTS
*For night rally
stages, Coopers
needed maximum
illumination. Straps
held on the headlight
stone-protectors.*

SPECIFICATIONS

MODEL Austin Mini Cooper (1963–69)

PRODUCTION 145,000 (all models)

BODY STYLE Saloon.

CONSTRUCTION All steel two-door
monocoque mounted on front and rear
sub-frames.

ENGINES Four-cylinder 970cc/
997cc/998cc/1071cc/1275cc.

POWER OUTPUT 65 bhp at 6500 rpm
to 76 bhp at 5800 rpm.

TRANSMISSION Four-speed,
no synchromesh on first.

SUSPENSION Independent front and
rear suspension with rubber cones and
wishbones (Hydrolastic from late 1964).

BRAKES Lockheed front discs with
rear drums.

MAXIMUM SPEED 161 km/h (100 mph)

0–60 MPH (0–96 KM/H) 12.9 sec

0–100 MPH (0–161 KM/H) 20 sec

A.F.C. 10.6 km/l (30 mpg)

INTERIOR
The Cooper has typical rally-car
features: wood-rim Moto-Lita wheel,
fire extinguisher, Halda trip meter, rev-
counter, stopwatches, and maplight.
The only features that would have been
standard equipment are the centre
speedo, heater, and switches.

AUSTIN-HEALEY *Sprite Mk1*

SOME AUTOMOTIVE ACADEMICS reckon all the best car designs have a recognizable face. If that is the case, few cars have a cuter one than this cheeky little fellow, with that ear-to-ear grinning grille and those wide-open, slightly astonished, eyes. Of course, it is those trademark bulging peepers that prompted the nickname "Frogeye", by which everyone now recognizes this engaging little character. So much of the Frogeye's character was borne of necessity. The Donald Healey Motor Company and Austin had already teamed up with the Austin-Healey 100. In 1958, its little brother, the Sprite, was born, a spartan sports car designed down to a price and based on the engine and running gear of the Austin A35 saloon, with a bit of Morris Minor too. Yet the Frogeye really was a sports car and had a sweet raspberry exhaust note to prove it.

DECEPTIVE LOOKS
At just under 3.5 m (11 ft 5 in), the Frogeye is not quite as small as it looks. Its pert looks were only part of the car's cult appeal, for with its firm, even, harsh, ride it had a traditional British sports car feel. A nimble performer, you could hustle it along a twisty road, cornering flat and snicking through the gears.

REAR VIEW
Perspex rear window offered only limited rear vision.

ENGINE

The Austin-Morris A-series engine was a little gem. It first appeared in the Austin A35 saloon and went on to power several generations of Mini *(see pages 44–47)*. In the Frogeye it was modified internally with extra strong valve springs and fitted with twin SU carburettors to give 50 bhp gross (43 bhp net). By today's standards it's no roadburner, but in the late Fifties it was a peppy little performer.

ENGINE ACCEESS
Rear-hinged alligator bonnet gives great engine access and makes the Frogeye a delight for DIY tinkerers.

LOW DOWN
The Frogeye's low stance aided flat cornering. Ground clearance was better than it looked: just under 12.7 cm (5 in).

BUMPERS
Bumpers with over-riders were a sensible and popular extra.

GVS 668

RACING LINKS

Sprites put up spirited performances at
Le Mans and Sebring in Florida, making club
racing affordable to ordinary enthusiasts.

THE FROG'S EYES

Donald Healey's original design incorporated
retracting headlamps like the later Lotus Elan
(see pages 344–45), but extra cost ruled these out.
As it was, the protruding headlamp pods
created a car with a character all of its own.
The complex one-piece bonnet in which the
lamps are set is made up of four main panels.

DUAL LIGHTS
*Sidelights doubled
as flashing
indicators.*

GVS 668

LATER INCARNATION

The design has a classic simplicity, free of needless chrome embellishment; there is no external door handle to interrupt the flowing flanks. In 1961 the Frogeye was re-clothed in a more conventional skin, and these follow-on Sprites, also badged as MG Midgets, lasted until 1979.

ROUND RUMP
It is not so much a boot, because it does not open; more a luggage locker with access behind the rear seats.

GEAR LEVER
Stubby gear lever was nicely positioned for the driver.

COSY COCKPIT

The Frogeye fits like a glove. Side curtains rather than wind-down windows gave some extra elbow room and everything is within reach for the sporting driver – speedo on the right and rev-counter on the left.

SPECIFICATIONS

MODEL Austin-Healey Sprite Mk1 (1958–61)

PRODUCTION 38,999

BODY STYLE Two-seater roadster.

CONSTRUCTION Unitary body/chassis.

ENGINE BMC A-Series 948cc, four-cylinder, overhead valve.

POWER OUTPUT 43 bhp at 5200 rpm.

TRANSMISSION Four-speed manual, synchromesh on top three ratios.

SUSPENSION *Front:* Independent, coil springs and wishbones; *Rear:* Quarter-elliptic leaf springs, rigid axle.

BRAKES Hydraulic, drums all round.

MAXIMUM SPEED 135 km/h (84 mph)

0–60 MPH (0–96 KM/H) 20.5 sec

A.F.C. 12.5–16 km/l (35–45 mpg)

AUSTIN-HEALEY *3000*

THE HEALEY HUNDRED WAS A sensation at the 1952 Earl's Court Motor Show. Austin's Leonard Lord had already contracted to supply the engines, but when he noticed the sports car's impact, he decided he wanted to build it too – it was transformed overnight into the Austin-Healey 100. Donald Healey had spotted a gap in the American sports car market between the Jaguar XK120 *(see pages 296–99)* and the cheap and cheerful MG T series *(see pages 366–69)*. His hunch was right, for about 80 per cent of all production went Stateside. Over the years this rugged bruiser became increasingly civilized. In 1956, it received a six-cylinder engine in place of the four, but in 1959 the 3000 was born. It became increasingly refined, with front disc brakes, then wind-up windows, and ever faster. Our featured car is the last of the line, a 3000 Mk3. Although perhaps verging on grand-tourer territory, it was still one of the fastest Big Healeys and has become a landmark British sports car.

WELL MATURED

The Austin-Healey put on weight over the years, became gradually more refined too, but stayed true to its original sports car spirit. It developed into a sure-footed thoroughbred sports car.

WHEELS AND WHITEWALLS
Wire wheels with knock-off hubs were options on some models, standard on others; whitewalls usually signify an American car.

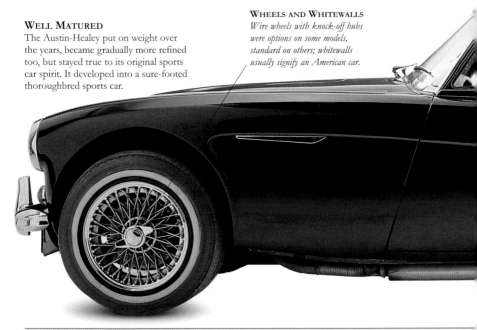

HOT PIT
Heat build-up from the engine and underfloor exhaust made for a warm ride.

COMFORTS
Updated weather equipment was an improvement on earlier efforts, which took two jugglers 10 minutes to erect.

BONNET SCOOP
All six-cylinder Healeys, both the 100/6 and the 3000, featured a bonnet scoop; the longer engine pushed the radiator forwards, with the scoop clearing the underbonnet protrusion to aid airflow.

WINDSCREEN
*In 1962, the 3000 acquired
a wrap-around windscreen
and wind-up windows, as the
once raw sports car adopted
trappings of sophistication.*

ENGINE

Under the bonnet of the biggest of the
so-called Big Healeys is the 2912cc straight-six,
designated the 3000. This is the butchest of the
big bangers, pumping out a hefty 150 bhp.

STYLING INFLUENCES

The two major influences on the Healey's
changing faces were the needs of the
American market and the impositions of
Austin, both as parts supplier and as frugal
keeper of purse-strings. But from the start,
the styling was always a major asset, and what
you see here in the 3000 Mk3 is the eventual
culmination of those combined styling forces.

REFINED REAR

The first prototype rear-end treatments
featured faddish fins that were replaced
by a classic round rump.

HEALEY GRIN

From the traditional Healey diamond grille, the mouth of the Austin-Healey developed into a wide grin.

INCREASED LUXURY

Once spartan, the cockpit of the Austin-Healey became increasingly luxurious, with a polished veneer fascia, lockable glove box, fine leather, and rich carpet. One thing remained traditional – engine heat meant the cockpit was always a hot place to be.

MORE POWER

The Americans bought more Healeys than anyone else and wanted more oomph. So in 1959 the 2639cc six-cylinder of the Healey 100/6 was bored out to 2912cc and rounded up to give the model name 3000.

SPECIFICATIONS

MODEL Austin-Healey 3000 (1959–68)

PRODUCTION 42,926 (all 3000 models)

BODY STYLES Two-seater roadster, 2+2 roadster, 2+2 convertible.

CONSTRUCTION Separate chassis/body.

ENGINE 2912cc overhead-valve, straight-six.

POWER OUTPUT 3000 Mk1: 124 bhp at 4600 rpm. 3000 Mk2: 132 bhp at 4750 rpm. 3000 Mk3: 150 bhp at 5250 rpm.

TRANSMISSION Four-speed manual with overdrive.

SUSPENSION *Front:* Independent coil springs and wishbones, anti-roll bar; *Rear:* Semi-elliptic leaf springs. Lever-arm dampers all round.

BRAKES Front discs; rear drum.

MAXIMUM SPEED 177–193 km/h (110–120 mph)

0–60 MPH (0–96 KM/H) 9.5–10.8 sec

A.F.C. 6–12 km/l (17–34 mpg)

BENTLEY *R-Type Continental*

IN ITS DAY THE BENTLEY CONTINENTAL, launched in 1952, was the fastest production four-seater in the world and acclaimed as "a modern magic carpet which annihilates distance". The R-Type Conti is still rightly considered one of the greatest cars of all time. Designed for the English country gentleman, it was understated, but had a lithe, sinewy beauty rarely seen in other cars of its era. Rolls-Royce's brief was to create a fast touring car for plutocrat customers, and to do that they had to reduce both size and weight. Aluminium construction helped the weight, while wind tunnel testing created that slippery shape. Those emergent fins at the back were not for decoration – they actually aided the car's directional stability. But such avant-garde development did not come cheap. In 1952 the R-Type Continental was the most expensive production car in the world and cost the same as a very large and agreeable house.

POST-WAR CLASSIC

In 1952, with wartime austerity a fading memory, this was one of the flashiest and most rakish cars money could buy. Today, this exemplar of breeding and privilege stands as a resplendent memorial to the affluence and optimism of Fifties Britain. Collectors seem to agree that the Continental is the finest post-war Bentley and one of the world's all-time great cars.

EXPORT ONLY

Such was the cost of the Continental that it was first introduced on an export-only basis.

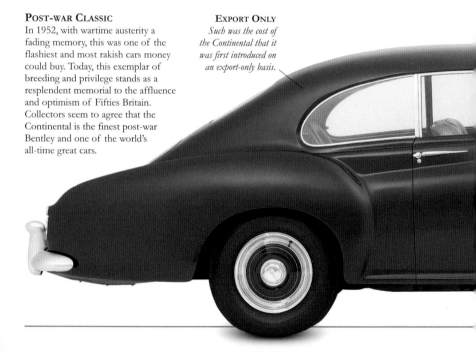

QUALITY RIDE
The Continental was a car that begged you to depress its accelerator pedal to the floor and reassured you with its powerful brakes.

DESIGN SIMILARITIES
The Continental bears an uncanny resemblance to a Pininfarina R-Type prototype shown at the 1948 Paris Salon.

RADIATOR
Classic Gothic radiator shell was considered far more sporting than Rolls-Royce's Doric example.

LIGHTS
Front fog lights used to be known as "pass lights" for overtaking.

UKL 109

AERODYNAMIC TESTS

The Continental spent much time in the wind tunnel to establish air drag during forward motion. Sweeping rear quarters directed the wind over the rear wheels, which were covered in spats to assist air flow. During prototype testing, it was found that a normal set of six-ply tyres lasted for only 32 km (20 miles).

REAR WINGS

Gently tapering rear wings funnelled air away into a slipstream; the Continental's aerodynamics were way ahead of its time.

ALUMINIUM CONSTRUCTION

Not only was the body made from lightweight aluminium – courtesy of H.J. Mulliner & Co. Ltd. – but also the side window and screen frames. The prototype had high quality alloy bumpers; production cars had steel ones.

CARBURETTORS

Carburation was by two SU HD8 units.

WEIGHT

Body weight was kept to a minimum because no Fifties' tyres could cope with speeds over 193 km/h (120 mph).

ENGINE

Continentals used a 4-litre straight-six engine of 4566cc – increased to 4887cc in May 1954 and known as the big bore engine. It allowed the car to reach 80 km/h (50 mph) in first gear.

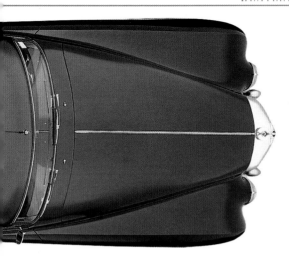

SPECIFICATIONS

MODEL Bentley R-Type Continental (1952–55)

PRODUCTION 208

BODY STYLE Two-door, four-seater touring saloon.

CONSTRUCTION Steel chassis, alloy body.

ENGINES 4566 or 4887cc straight-sixes.

POWER OUTPUT Never declared, described as "sufficient".

TRANSMISSION Four-speed synchromesh manual or auto option.

SUSPENSION Independent front with wishbones and coil springs, rear live axle with leaf springs.

BRAKES Front disc, rear drums.

MAXIMUM SPEED 185 km/h (115 mph)

0–60 MPH (0–96 KM/H) 13.5 sec

0–100 MPH (0–161 KM/H) 36.2 sec

A.F.C. 6.9 km/l (19.4 mpg)

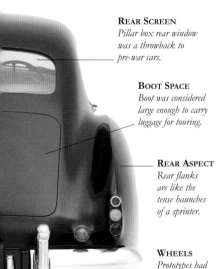

REAR SCREEN
Pillar box rear window was a throwback to pre-war cars.

BOOT SPACE
Boot was considered large enough to carry luggage for touring.

REAR ASPECT
Rear flanks are like the tense haunches of a sprinter.

WHEELS
Prototypes had spats covering the rear wheels.

PLUSH DASH
The beautifully detailed dashboard mirrored the Continental's exterior elegance. The first R-Types had manual gearboxes with a right-hand floor-mounted lever, thus reflecting the car's sporting character. Later models were offered with automatic boxes.

BENTLEY *Flying Spur*

ARGUABLY THE MOST BEAUTIFUL post-war Bentley, the Flying Spur was the first four-door Continental. Initially, Rolls-Royce would not allow coachbuilder H.J. Mulliner to use the name Continental, insisting it should only apply to two-door cars. After months of pressure from Mulliner, R.R. relented and allowed the shapely coach-built car to be known as a proper Continental. More than worthy of the hallowed name, the Flying Spur was launched in 1957, using the standard S1 chassis. In 1959 it inherited R.R.'s 220 bhp, oversquare, light-alloy V8, and by July 1962 the bodyshell was given the double headlamp treatment and upgraded into what some consider to be the best of the breed – the S3 Flying Spur. Subtle, understated, and elegant, Flying Spurs are rare, and in their day were among the most admired and refined machines in the world. Although sharing much with the contemporary Standard Steel Bentley, the Spur's list price was half as much more as the stock item.

INTERIOR
Interior includes carefully detailed switchgear, the finest hide and walnut, and West of England cloth.

WEIGHTY REAR
Tapering tail and swooping roof line managed to lend an air of performance.

POWER STEERING
The large, spindly steering wheel was power-assisted.

HAND-BUILT REFINEMENTS

Coachbuilder H.J. Mulliner would receive the chassis from Rolls-Royce and clothe it with a hand-built body. Although customers would often have to wait up to 18 months for their cars to be completed, the finished product was considered the zenith of good taste and refinement.

ENGINE
V8 had aluminium cylinder heads, block, and pistons.

SPECIFICATIONS

MODEL S3 Bentley Continental H.J. Mulliner Flying Spur (1962–66)

PRODUCTION 291

BODY STYLE Four-door, five-seater.

CONSTRUCTION Aluminium body, separate steel cross-braced box section chassis.

ENGINE V8, 6230cc.

POWER OUTPUT Never officially declared.

TRANSMISSION Four-speed automatic.

SUSPENSION *Front:* independent coil springs and wishbones; *Rear:* semi-elliptic leaf springs.

BRAKES Four-wheel Girling drums.

MAXIMUM SPEED 185 km/h (115 mph)

0–60 MPH (0–96 KM/H) 10.8 sec

0–100 MPH (0–161 KM/H) 34.2 sec

A.F.C. 4.9 km/l (13.8 mpg)

FRONT ASPECT

The Flying Spur's four-headlamp nose was shared with the standard steel Bentley S3, along with a lowered radiator and bonnet line. The body was constructed from hand-rolled aluminium.

BENTLEY *Continental Supersports*

THE 2003 CONTINENTAL GT with its magnificent W12 engine changed Bentley forever. Compact, rapid, reliable, and fashionable, the Conti (to use its street name) is one of the Crewe firm's most admired products and brought the Bentley brand to a younger customer. In 2009, the Supersports was launched as the fastest production Bentley ever, and the first to run on petrol and biofuel (E85 ethanol). A special Quickshift tiptronic six-speed automatic gearbox reduced gear change times by 50 per cent, and a Torsen T3 four-wheel-drive system made the Supersports sure-footed enough to break the world speed record on ice at 330 km/h (205 mph). But despite the epic performance this is an amazingly refined supercar with superb steering and a truly magic carpet ride.

SIT LIGHTLY
Seat frames in the Supersports are carbon fibre to save weight.

THE PEOPLE'S POWERPLANT
The VW-designed twin-turbocharged six-litre W12 engine is one of the world's best. The cylinders are arranged in four banks of three to save space, but even in standard GT tune this engine still develops a mighty 552 bhp.

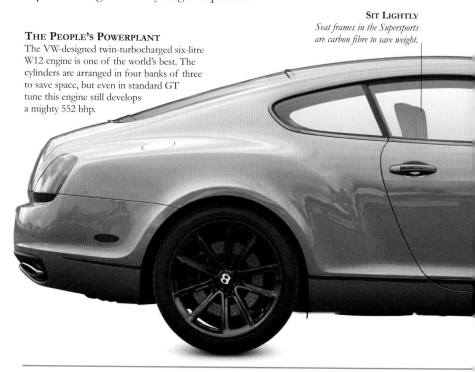

KEEPING COOL
Huge grille and vents cool engine and brakes.

SUPERSPORTS CLONES
The superfast Bentley has become so iconic and desirable that owners of "ordinary" Conti GTs often put on the revised bumpers and black rims to make their cars look like Supersports. There's now a flourishing global industry "pimping" up all Continentals.

MONSTER BRAKES
Ceramic brakes are the most powerful ever fitted to a production car.

NOSE JOB
Front apron is unique to Supersports and is plastic to save weight.

BMW *507*

WHOEVER WOULD HAVE THOUGHT that in the mid-Fifties BMW would have unveiled something as voluptuously beautiful as the 507? The company had a fine pre-World War II heritage that culminated in the crisp 328, but it did not resume car manufacturing until 1952, with the curvy, but slightly plump, six-cylinder 501 saloon. Then, at the Frankfurt show of late 1955 they hit us with the 507, designed by Count Albrecht Goertz. The 507 was a fantasy made real; not flashy, but dramatic and with poise and presence. BMW hoped the 507 would straighten out its precarious finances, winning sales in the lucrative American market. But the BMW's exotic looks and performance were more than matched by an orbital price. Production, which had been largely by hand, ended in March 1959 after just 252 – some say 253 – had been built. In fact, the 507 took BMW to the brink of financial oblivion, yet if that had been the last BMW it would have been a beautiful way to die.

TEUTONIC LINKS
Mounted on a tubular-steel chassis cut down from saloons, Albrecht Goertz's aluminium body is reminiscent of the contemporary – and slightly cheaper – Mercedes-Benz 300SL roadster; from the front it resembles the later AC Aces and Cobra *(see pages 12–19).*

BRAKES
Most 507s were built with all-round Alfin drum brakes. Some later cars had more effective front disc brakes.

DOOR HANDLES
Like the bumpers, the door handles were surprisingly discreet – if not particularly easy to use.

TOOLKIT
Like all modern BMW's, the 507 had a toolkit in the boot.

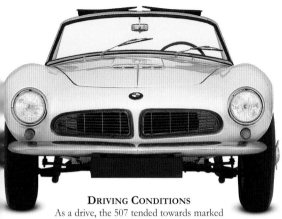

SPECIFICATIONS

MODEL BMW 507 (1956–59)

PRODUCTION 252/3, most LHD

BODY STYLE Two-seater roadster.

CONSTRUCTION Box section and tubular steel chassis; aluminium body.

ENGINE All-aluminium 3168cc V8, two valves per cylinder.

POWER OUTPUT 150 bhp at 5000 rpm; some later cars 160 bhp at 5600 rpm.

TRANSMISSION Four-speed manual.

SUSPENSION *Front:* Unequal-length wishbones, torsion-bar springs and telescopic dampers; *Rear:* Live axle, torsion-bar springs.

BRAKES Drums front and rear; front discs and rear drums on later cars.

MAXIMUM SPEED 201 km/h (125 mph); 217–225 km/h (135–140 mph) with optional 3.42:1 final drive.

0–60 MPH (0–96 KM/H) 9 sec

A.F.C. 6.4 km/l (18 mpg)

DRIVING CONDITIONS
As a drive, the 507 tended towards marked understeer; so instant was throttle response that the tail easily snapped out.

BONNET VENT
Ornate chrome-plated grilles in the front wings covered functional engine-bay air vents.

OPTIONAL POWER
Tuned 160 bhp versions of the 507 were good for 225 km/h (140 mph).

WHEELS
Knock-off Rudge wheels like this were the most sought-after option.

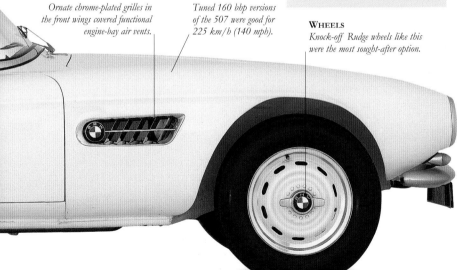

RAKISH BODY

The 507's body is an all-aluminium affair atop a simple tubular chassis. Brightwork is kept to the minimum, accentuating the clean lines. The brightwork included on the car is kept simple; the rear bumpers, for example, have no bulky over-riders.

ENGINE

The 3.2-litre all-aluminium engine was light and powerful. Twin Zenith carbs are the same as those of the contemporary Porsches.

HOOD

You rarely see a 507 with its hood raised, but it is simple to erect and remarkably handsome.

PIPE MUSIC

The BMW had a brisk, wholesome bark and unmistakable creamy wuffle of a V8.

BEEMER BADGING

Eight BMW stylized propeller roundels, including those on wheel trims and eared spinners, grace the 507, nine if you include the badge in the centre of the steering wheel.

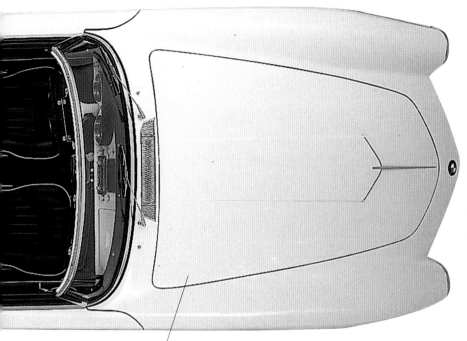

ENGINE PROBLEMS
*The 3.2-litre engine tended
to run too hot in traffic and
too cool on the open road.*

HORN-PULLS
*The interior was clearly inspired
by US styling of the period,
with gimmicky horn-pulls
behind the steering wheel.*

INTERIOR

The 507, unlike the contemporary 503, has
a floor-mounted lever to operate the four-
speed gearbox. Dash consists of a clock,
speedometer, and rev-counter. Some cars
had internally adjustable door mirrors.

BMW *3.0CSL*

ONE LITTLE LETTER CAN MAKE SO much difference. In this case it is the L at the
end of the name tag that makes the BMW 3.0CSL so special. The BMW CS
pillarless coupés of the late Sixties and early Seventies were elegant and good-
looking glasshouse tourers. But add that L and you have a legend. The letter
actually stands for "Leichtmetall", and when tacked to the rump of the BMW it
amounts to warpaint. The original CSL of 1974 had a 2985cc engine developing
180 bhp, no front bumper, and a mixture of aluminium and thin steel body-panels.
In August 1972, a cylinder-bore increase took the CSL's capacity to 3003cc with
200 bhp and allowed it into Group 2 for competition purposes. But it is the wild-
winged, so-called "Batmobile" homologation special that really boils the blood
of boy racers. An ultimate road car, great racer, rare, short-lived and high-priced,
this charismatic, pared-down Beemer has got copy book classic credentials.

GOOD LOOKER

Even mild rather than wild and winged, the
CSL is certainly one of the best-looking
cars of its generation. With its pillarless
look, the cabin is light and airy, despite the
black interior. But all that glass made it hot;
air vents behind the BMW rear-pillar
badge helped a little.

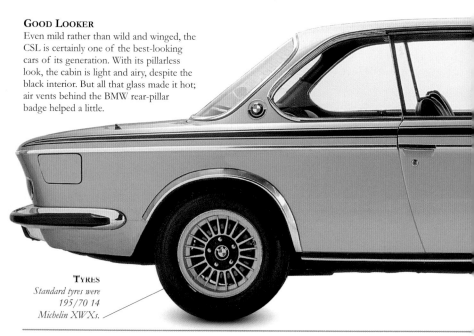

TYRES
*Standard tyres were
195/70 14
Michelin XWXs.*

RACING TRIM
Optional air guide for rear end of roof was available, along with seven other aerodynamic aids.

STEERING WHEEL
Steering wheel was straight out of the CS/CSi.

CALLING CARD
Large script leaves no one in any doubt about what has just overtaken them.

BOOT
The first CSLs came with aluminium boot, bonnet, and doors.

KS·K 5405

BODY PANELS
"Leichtmetall" meant body panels were made of aluminium and thinner-than-standard steel.

SPOILER EXCESS
For homologation purposes, at least 500 road cars had to be fitted with a massive rear spoiler – it was considered so outrageous that most were supplied for owners to fit at their discretion.

RACING UNIT
The CSL's race engine grew from 3.2 to 3.5 litres.

3.0 CSL

BRAKES
Ventilated discs were necessary to counter the CSL's immense power.

ENGINE

In genuine racing trim, the Batmobile's 3.2-litre straight-six engine gave nearly 400 bhp and, for 1976, nearly 500 bhp with turbocharging. But road cars like this British-spec 3003cc 3.0CSL gave around 200 bhp on fuel injection.

SEVENTIES' BARGAIN
Just after the 1973 fuel crisis, you could pick up a CSL for very little money.

ENGINE UPGRADE
Early CSLs had a carburettor-fed 2985cc engine developing 180 bhp; after 1972, capacity increased to 3003cc, shown here, for homologation purposes.

BUMPER-TO-BUMPER
German-market CSLs had no front bumper and a glass-fibre rear bumper; this car's metal items show it to be a British-spec model.

TRACK SUCCESS

The CSLs were the first BMWs developed under the company's new Motorsport department which was set up in 1972. The model produced immediate success for BMW, initially in Europe and then on tracks in the United States. The CSL won all but one of the European Touring Car Championships between 1973 and 1979.

INTERIOR
British-spec CSLs, like this car, retained Scheel lightweight bucket seats, but had carpets, electric windows (front and rear), power steering, and a sliver of wood.

LIMITED EDITION
500 fuel-injected versions of the CSL were offered by British tuners.

SPECIFICATIONS

MODEL BMW 3.0CSL (1971–74)

PRODUCTION 1,208 (all versions)

BODY STYLE Four-seater coupé.

CONSTRUCTION Steel monocoque, steel and aluminium body.

ENGINES 2985cc, 3003cc, or 3153cc inline six.

POWER OUTPUT 200 bhp at 5500 rpm (3003cc).

TRANSMISSION Four-speed manual.

SUSPENSION *Front:* MacPherson struts and anti-roll bar; *Rear:* semi-trailing swinging arms, coil springs, and anti-roll bar.

BRAKES Servo-assisted ventilated discs front and rear.

MAXIMUM SPEED 217 km/h (135 mph) (3003cc)

0–60 MPH (0–96 KM/H) 7.3 sec (3003cc)

0–100 MPH (0–161 KM/H) 21 sec (3003cc)

A.F.C. 7.8–8.8 km/l (22–25 mpg)

DO-IT-YOURSELF
Road-going cars were only slightly lighter than the CS and CSi; they even had BMW's trademark toolkit, neatly hinged from the underside of the boot lid.

BMW *M1*

THE M1 – A SIMPLE NAME, a simple concept. M stood for Motorsport GmbH, BMW's separate competition division, and the number one? Well, this was going to be a first, for this time BMW were not just going to develop capable racers from competent saloons and coupés. They were going to build a high-profile, beat-all racer, with road-going versions basking in the reflected glory of on-track success. The first prototype ran in 1977, with the M1 entering production in 1978. By the end of manufacture in 1980, a mere 457 racing and road-going M1s had been built, making it one of the rarest and most desirable of modern BMWs. Though its racing career was only briefly distinguished, it is as one of the all-time ultimate road cars that the M1 stands out, for it is not just a 257 km/h (160 mph) "autobahnstormer". It is one of the least demanding supercars to drive, a testament to its fine engineering, and is in many ways as remarkable as the gorgeous 328 of the 1930s.

INTERNATIONAL CONSTRUCTION
The M1 had widespread international influences. From a concept car created in 1972 by Frenchman Paul Bracq, the final body shape was created in Italy by Giorgio Giugiaro's ItalDesign in Turin. Lamborghini also contributed to the engineering. Yet somehow it all comes together in a unified shape and, with the double kidney grille, the M1 is still unmistakably a BMW.

SUSPENSION
Suspension was a mix of springs, wishbones, and telescopic dampers.

LEFT HOOKERS
All BMW M1s were left-hand drive.

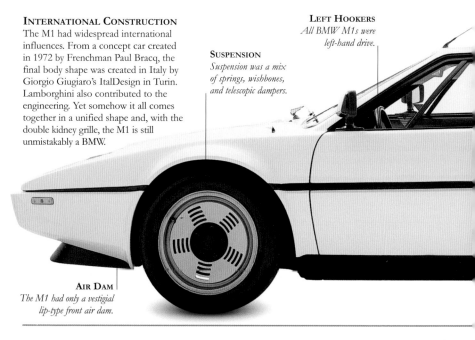

AIR DAM
The M1 had only a vestigial lip-type front air dam.

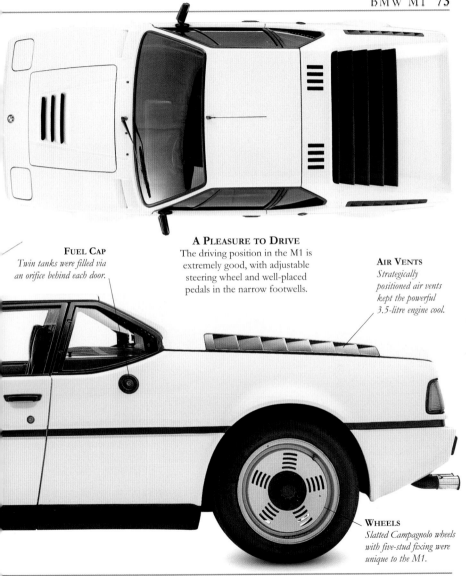

FUEL CAP
*Twin tanks were filled via
an orifice behind each door.*

A PLEASURE TO DRIVE
The driving position in the M1 is
extremely good, with adjustable
steering wheel and well-placed
pedals in the narrow footwells.

AIR VENTS
*Strategically
positioned air vents
kept the powerful
3.5-litre engine cool.*

WHEELS
*Slatted Campagnolo wheels
with five-stud fixing were
unique to the M1.*

CYLINDER HEAD
*The cylinder head was a
light-alloy casting, with
two chain-driven
overhead cams operating
four valves per cylinder.*

ENGINE
The M1's 3453cc straight-six
engine uses essentially the
same cast-iron cylinder block
as BMW's 635CSi coupé,
but with a forged-alloy
crankshaft and slightly
longer connecting rods.

MIRRORS
*Big door mirrors –
essential for manoeuvring
the M1 – were
electrically adjusted.*

MZ-X 498

DARK INTERIOR

The all-black interior is sombre, but fixtures and fittings are all to a high standard; unlike those of many supercars, the heating and ventilation systems actually work. However, rearward visibility through the slatted, heavily buttressed engine cover is severely restricted.

REAR LIGHTS

Large rear lamp clusters were the same as those of the 6-series coupé and 7-series saloon models.

HEADLAMPS

Retractable headlamps were backed up by grille-mounted driving lights.

SPECIFICATIONS

MODEL BMW M1 (1978–80)

PRODUCTION 457, all LHD

BODY STYLE Two-seater mid-engined sports.

CONSTRUCTION Tubular steel space-frame with glass-fibre body.

ENGINE Inline six, four valves per cylinder, dohc 3453cc.

POWER OUTPUT 277 bhp at 6500 rpm.

TRANSMISSION Combined ZF five-speed gearbox and limited slip differential.

SUSPENSION Coil springs, wishbones, and Bilstein gas-pressure telescopic dampers front and rear.

BRAKES Servo-assisted ventilated discs all round.

MAXIMUM SPEED 261 km/h (162 mph)

0–60 MPH (0–96 KM/H) 5.4 sec

A.F.C. 8.5–10.6 km/l (24–30 mpg)

PURE M1 RACING

BMW teamed up with FOCA (Formula One Constructors' Association) to create the Procar series – M1-only races planned primarily as supporting events for Grand Prix meetings in 1979 and 1980.

BUGATTI *Veyron Grand Sport*

THE VEYRON IS QUITE SIMPLY the greatest car ever made. This isn't just the fastest production car in the world, it's a technological *tour de force* that defies physics, gravity, and common sense. With a 0–62 mph (0–100 km/h) time of less than 2.7 seconds, the acceleration of the Veyron's open-top version, the Grand Sport, has been described as "like falling out of an aeroplane", yet its road manners are the pinnacle of civility. The legendary 16-cylinder, quad-turbocharged engine develops 1,001 bhp, yet it can brake from 100 km/h (62 mph) to a standstill in only 31.4 m (103 ft). The Veyron is more than just a million-pound supercar, it's a global celebrity in its own right.

BUGATTI REINVENTED
When VW bought the Bugatti brand in 1998 nobody could have guessed their plans for such a stunning reinvention, or that they would create the world's most audacious supercar so quickly. In 1999, they surprised the world by showing the first concept, and by 2005 had created an automotive masterpiece that will never be repeated in our lifetime.

TARGA TOP
Grand Sport versions have a removable, lightweight, transparent polycarbonate roof.

WING BRAKE
Air brake rises in 0.4 seconds and has the stopping power of an ordinary hatchback.

SPECIFICATIONS

MODEL Bugatti Veyron 16.4 Grand Sport (2008)

PRODUCTION 450 (all variants)

BODY STYLE Two-door, two-seater open-top sports car.

CONSTRUCTION Carbon-fibre composite and alloy.

ENGINE 7,993cc W16.

POWER OUTPUT 1,001 bhp.

TRANSMISSION Seven-speed, sequential, dual-clutch automatic.

SUSPENSION Computer-controlled double wishbones front and rear.

BRAKES Carbon-fibre, silicone carbide discs plus rear airbrake.

MAXIMUM SPEED 407 km/h (253 mph)

0–62 MPH (0–100 KM/H) less than 2.7 sec

A.F.C. 4 km/l (9.5 mpg)

EPIC GEARBOX

Steering wheel has magnesium gearbox paddles for the seven-speed direct-shift, dual-clutch, computer-controlled automatic made by Ricardo in England. Shift speed is 150 milliseconds and the car's torque is an incredible 1,250 Nm at 2,200–5,500 rpm.

OPEN AIR
With roof removed, speeds of up to 350km/h (217 mph) are possible.

BRILLIANT BRAKES
Brakes are cross-drilled, radially vented silicon carbide discs.

COOL DOWN
Seven separate radiators keep engine, gearbox, and differential cool.

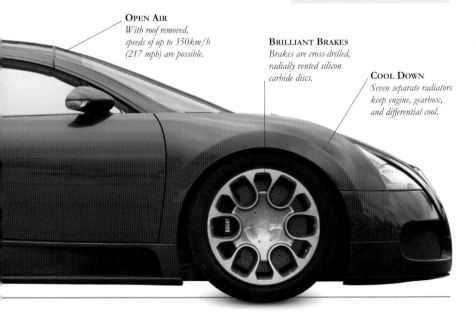

BUICK *Roadmaster (1949)*

THE '49 ROADMASTER TOOK THE market's breath away. With a low silhouette, straight bonnet, and fastback styling, it was a poem in steel. The first Buick with a truly new post-war look, the '49 was designed by Ned Nickles using GM's new C-body. It also boasted two bold new styling motifs: Ventiports and an aggressive 25-tooth "Dollar Grin" grille. Harley Earl's aesthetic of aeronautical entertainment worked a treat and Buick notched up nearly 400,000 sales that year. Never mind that the windscreen was still two-piece, that there was no power steering, and the engine was a straight-eight – it looked gorgeous and came with the new Dynaflow automatic transmission. The Roadmaster, like the '49 Cadillac, was a seminal car and the first flowering of the most flamboyant decade of car design ever seen.

SERIOUS CACHET
For years GM's copywriters crowed that "when better cars are built, Buick will build them", and in a sense that hyperbole was true. In its day, the gloriously voluptuous Roadmaster was a serious set of wheels, only one step down from a Cadillac, and to own one meant you really had arrived.

SPOTLIGHT
Spotlight with mirror was a $25 option.

VENTIPORT STATUS
Cheaper Buicks had only three Ventiports; the lavish Roadmaster had four.

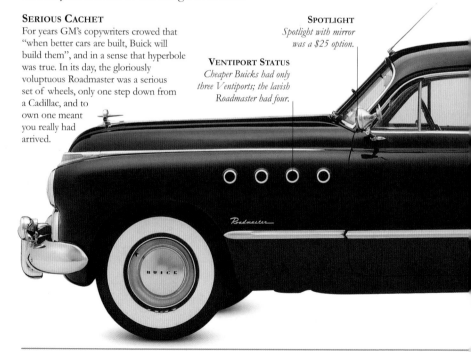

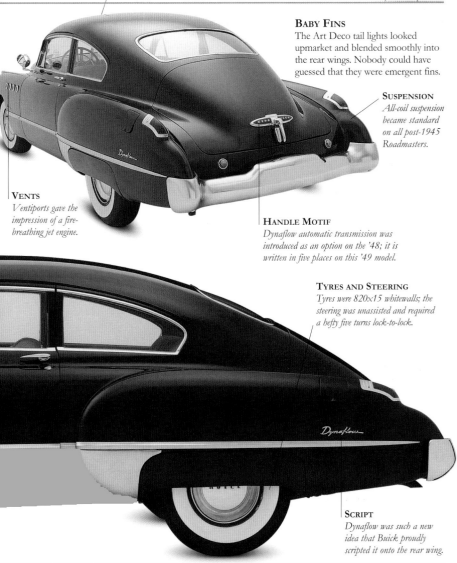

BABY FINS
The Art Deco tail lights looked upmarket and blended smoothly into the rear wings. Nobody could have guessed that they were emergent fins.

SUSPENSION
All-coil suspension became standard on all post-1945 Roadmasters.

VENTS
Ventiports gave the impression of a fire-breathing jet engine.

HANDLE MOTIF
Dynaflow automatic transmission was introduced as an option on the '48; it is written in five places on this '49 model.

TYRES AND STEERING
Tyres were 820x15 whitewalls; the steering was unassisted and required a hefty five turns lock-to-lock.

SCRIPT
Dynaflow was such a new idea that Buick proudly scripted it onto the rear wing.

ADVERTISING
The '49's class set the trend for later Roadmasters, with the copywriters eager to stress that the model was the "Buick of Buicks".

Sovereign right of a successful man

Roadmaster

SIGN OF THE TIMES
The Roadmaster may have shared its body with the Oldsmobile 98 and the Cadillac Series 62, but it gave Buick a distinction never seen before. Big, bold, and brash, the '49 was perfect for its time and it began the trend for lower, sleeker styling. Optimistic, opulent, and glitzy, it carried strident styling cues that told people a block away that this was no ordinary car, this was a Buick – even better, the very best Buick money could buy.

CLASSY REAR
Elegant flourish completed the swooping teardrop rear.

DYNA FLOW

18 363
52 WYOMING

STYLING
The GM C-body had closed quarters and Sedanette styling.

EARLY TRADEMARKS
Gun-sight bonnet ornament, bucktooth grille, and Ventiports were flashy styling metaphors that would become famous Buick trademarks. Although divided by a centre pillar, the windscreen glass was actually curved.

ENGINE
The Roadie had a Fireball straight-eight cast-iron 320cid engine.

GRILLE
The classic vertical grille bars were replaced for the 1955 model year.

DASHBOARD
The instrument panel was new for '49 and described as "pilot centred" because the speedo was positioned straight ahead of the driver through the steering wheel.

SPECIFICATIONS

MODEL 1949 Buick Roadmaster Series 70 Sedanette
PRODUCTION 18,415 (1949)
BODY STYLE Two-door fastback coupé.
CONSTRUCTION Steel body and chassis.
ENGINE 320cid straight-eight.
POWER OUTPUT 150 bhp.
TRANSMISSION Two-speed Dynaflow automatic.
SUSPENSION Front and rear coil springs.
BRAKES Front and rear drums.
MAXIMUM SPEED 161 km/h (100 mph)
0–60 MPH (0–96 KM/H) 17 sec
A.F.C. 7 km/l (20 mpg)

BUICK *Roadmaster (1957)*

IN 1957, AMERICA WAS GEARING up for the Sixties. Little Richard screamed his way to the top with "Lucille" and Elvis had nine hits in a row. Jack Kerouac penned his immortal novel *On the Road*, inspiring carloads of Americans to seek the adman's "Promised Land" along Ike's new interstates. Fins and chrome were applied with a trowel and General Motors spent several hundred million dollars refashioning their Buick model range. The Roadmaster of 1957 was low and mighty, a massive 5.46 m (17 ft 11 in) long and 1.83 m (6 ft) wide. Power was up to 300 bhp, along with trendy dorsal fins, sweepspear body mouldings, and a trio of chrome chevrons on the rear quarters. Four Ventiports, a Buick trademark harking back to the original 1949 Roadmaster, still graced the sweeping front wings. But America did not take to Buick's new look, particularly some of the Roadmaster's fashionable jet-age design motifs.

PLANE STYLING
Aircraft design exerted a big influence on automotive styling in the Fifties and the '57 Roadmaster was no exception. With wrap-around windscreen, cockpit-like roof area, and turbine-style wheel covers, a nation of Walter Mittys could imagine themselves vapour-trailing through the stratosphere.

CABIN OR COCKPIT?
Rakish swooping roof line borrowed heavily from bubble cockpits of jet fighters.

HEIGHT
The '57 Roadmaster was lower and sleeker than previous models.

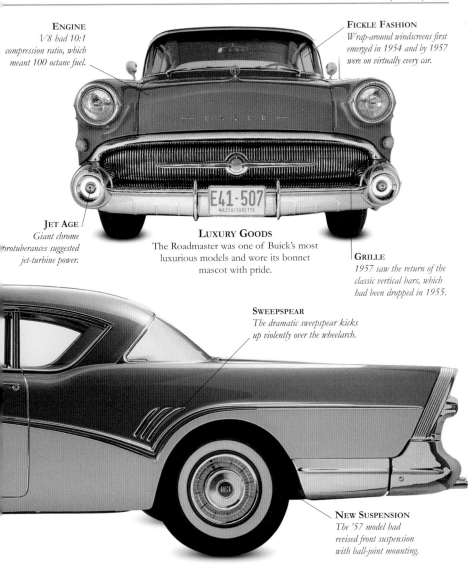

ENGINE
V8 had 10:1 compression ratio, which meant 100 octane fuel.

FICKLE FASHION
Wrap-around windscreens first emerged in 1954 and by 1957 were on virtually every car.

JET AGE
Giant chrome protuberances suggested jet-turbine power.

LUXURY GOODS
The Roadmaster was one of Buick's most luxurious models and wore its bonnet mascot with pride.

GRILLE
1957 saw the return of the classic vertical bars, which had been dropped in 1955.

SWEEPSPEAR
The dramatic sweepspear kicks up violently over the wheelarch.

NEW SUSPENSION
The '57 model had revised front suspension with ball-joint mounting.

E41-507
MASSACHUSETTS

ENGINE

The hot Buick's 5.9-litre V8 pushed out 300 bhp; it was capable of 180 km/h (112 mph) and 0 to 60 mph (96 km/h) in 10 seconds. Dynaflow transmission had variable pitch blades which changed their angle like those of an aeroplane propeller.

BOOT

The cavernous boot could accommodate almost anything you could buy at the mall in the consumer-driven Fifties.

STYLING EXCESS

Vast chrome rear bumper made for a prodigious overhang, with massive Dagmar-like over-riders, razor-sharp tail lights and fluted underpanel – a stylistic nightmare. One interesting new feature was the fuel filler-cap, now positioned in a slot above the rear number plate.

FIN DETAIL

The Roadmaster showed that, by 1957, tail-fin fashion was rising to ridiculous heights. Unfortunately, by '57 the Roadmaster looked very much like every other American car. Gone was that chaste individuality, and Buick began to lose its reputation as a maker of high-quality cars – production fell by 24 per cent this year.

LIMITED VISION
Small tinted rear windscreen didn't offer much assistance to the driver in reversing situations.

SPECIFICATIONS

MODEL Buick Roadmaster (1957)
PRODUCTION 36,638 (1957)
BODY STYLE Two-door, five-seater hardtop coupé.
CONSTRUCTION X-braced chassis with steel body.
ENGINE V8, 364cid.
POWER OUTPUT 250 bhp at 4400 rpm.
TRANSMISSION Dynaflow two-speed automatic.
SUSPENSION Independent coil springs.
BRAKES Hydraulic servo drums all round.
MAXIMUM SPEED 180 km/h (112 mph)
0–60 MPH (0–96 KM/H) 10.5 sec
0–100 MPH (0–161 KM/H) 21.2 sec
A.F.C. 4.2 km/l (12 mpg)

POWER STEERING
Power-assisted steering and Dynaflow automatic transmission became standard on all Roadmasters from 1953.

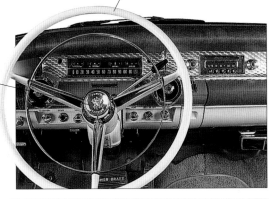

GM BADGING
Badge at the centre of the steering wheel indicates that Buicks were built at GM's factory in Flint, Michigan.

INTERIOR
Roadmaster standard special equipment included a Red Liner speedometer, glovebox lamp, trip mileage indicator, and a colour-coordinated dash panel. From 1955 Roadmasters could be ordered with a choice of 10 types of interior trim.

Buick *Limited Riviera (1958)*

When your fortunes are flagging, you pour on the chrome. As blubbery barges go, the '58 Limited has to be one of the gaudiest. Spanning 5.78 m (19 ft) and tipping the scales at two tonnes, the Limited is empirical proof that 1958 was not Buick's happiest year. Despite all that twinkling kitsch and the reincarnated Limited badge, the bulbous Buick bombed. For a start, GM's Dynaflow automatic transmission was not up to Hydra-Matic standards, and the Limited's brakes were disinclined to work. Furthermore, in what was a recession year for the industry, the Limited had been priced into Cadillac territory – $33 more than the Series 62. Total production for the Limited in 1958 was a very limited 7,436 units. By the late Fifties, Detroit had lost its way, and the '58 Limited was on the road to nowhere.

TRIMMINGS
Interiors were trimmed in grey cloth and vinyl or Cordaveen. Seat cushions had Double-Depth foam rubber.

CHILD OF THE FIFTIES
Buick's answer to an aircraft carrier was a riot of ornamentation that went on for half a block. At rest, the Limited looked like it needed a fifth wheel to support that weighty rear overhang.

BODY STYLES
As well as this four-door Riviera, the 700 Series also included a two-door version, a stripped chassis model, and a convertible.

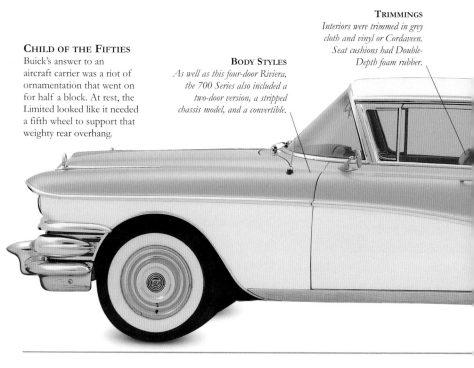

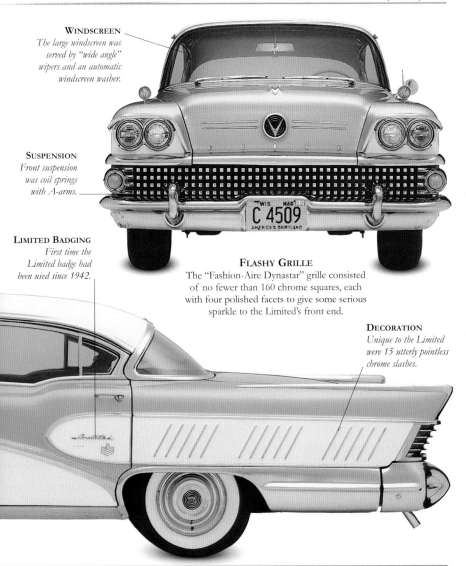

WINDSCREEN
The large windscreen was served by "wide angle" wipers and an automatic windscreen washer.

SUSPENSION
Front suspension was coil springs with A-arms.

LIMITED BADGING
First time the Limited badge had been used since 1942.

FLASHY GRILLE
The "Fashion-Aire Dynastar" grille consisted of no fewer than 160 chrome squares, each with four polished facets to give some serious sparkle to the Limited's front end.

DECORATION
Unique to the Limited were 15 utterly pointless chrome slashes.

C 4509
WIS MAR 58
AMERICA'S DAIRYLAND

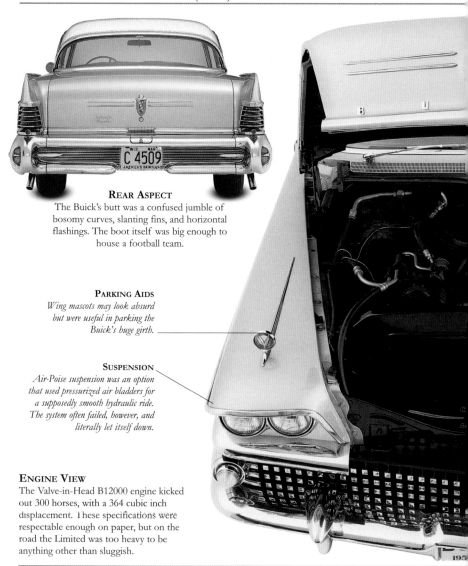

REAR ASPECT

The Buick's butt was a confused jumble of bosomy curves, slanting fins, and horizontal flashings. The boot itself was big enough to house a football team.

PARKING AIDS

Wing mascots may look absurd but were useful in parking the Buick's huge girth.

SUSPENSION

Air-Poise suspension was an option that used pressurized air bladders for a supposedly smooth hydraulic ride. The system often failed, however, and literally let itself down.

ENGINE VIEW

The Valve-in-Head B12000 engine kicked out 300 horses, with a 364 cubic inch displacement. These specifications were respectable enough on paper, but on the road the Limited was too heavy to be anything other than sluggish.

195

CHROME TRIM
The metal with a shiny coating could be found on everything from food mixers to radios in the Fifties.

ECONOMY
Producing 4.6 km/l (13 mpg), the Limited was thirsty.

SPECIFICATIONS

MODEL Buick Limited Riviera Series 700 (1958)

PRODUCTION 7,436 (1958, all Series 700 body styles)

BODY STYLES Two- and four-door, six-seater hardtops, two-door convertible.

CONSTRUCTION Steel monocoque.

ENGINE 364cid V8.

POWER OUTPUT 300 bhp.

TRANSMISSION Flight-Pitch Dynaflow automatic.

SUSPENSION *Front:* coil springs with A-arms; *Rear:* live axle with coil springs. Optional air suspension.

BRAKES Front and rear drums.

MAXIMUM SPEED 177 km/h (110 mph)

0–60 MPH (0–96 KM/H) 9.5 sec

A.F.C. 4.6 km/l (13 mpg)

HORN
Horn-pulls were pretty much standard on every US car in the Fifties.

INTERIOR
Power steering and brakes were essential and came as standard. Other standard equipment included an electric clock, cigarette lighters, and electric windows.

BUICK *Riviera (1964)*

IN '58, SO THE STORY GOES, GM's design supremo Bill Mitchell was entranced by a Rolls-Royce he saw hissing past a London hotel. "What we want", said Mitchell, "is a cross between a Ferrari and a Rolls". By August 1960, he'd turned his vision into a full-size clay mock-up. One of the world's most handsome cars, the original '63 Riviera locked horns with Ford's T-Bird and was GM's attempt at a "Great New American Classic Car". And it worked. Separate and elegant, the Riv was a clever amalgam of razor edges and chaste curves, embellished by just the right amount of chrome. Beneath the exquisite lines was a cross-member frame, a 401cid V8, power brakes, and a two-speed Turbine Drive tranny. In the interests of exclusivity, Buick agreed that only 40,000 would be made each year. With ravishing looks, prodigious performance, and the classiest image in town, the Riv ranks as one of Detroit's finest confections.

DIMENSIONS
Relatively compact, the Riviera was considerably shorter and lighter than other big Buicks.

CLASS APPEAL
The Riv was America's answer to the Bentley Continental, and pandered to Ivy League America's obsession with aristocratic European thoroughbreds like Aston Martin, Maserati, and Jaguar.

FINE-LINE DESIGN
Superbly understated, razor-edged styling made for an unfussy, crisp-looking machine.

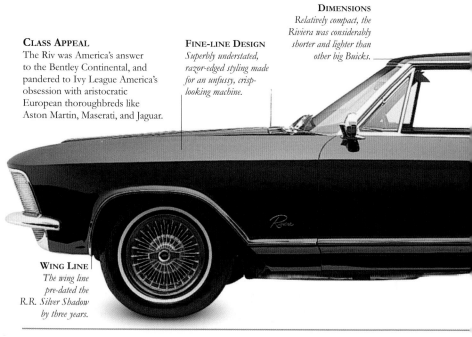

WING LINE
The wing line pre-dated the R.R. Silver Shadow by three years.

SPECIFICATIONS

MODEL Buick Riviera (1964)
PRODUCTION 37,958 (1964)
BODY STYLE Two-door hardtop coupé.
CONSTRUCTION Steel body and chassis.
ENGINE 425cid V8
POWER OUTPUT 340–360 bhp.
TRANSMISSION Two- or three-speed automatic.
SUSPENSION Front and rear coil springs.
BRAKES Front and rear drums.
MAXIMUM SPEED 193–201 km/h (120–125 mph)
0–60 MPH (0–96 KM/H) 8 sec
A.F.C. 4.2–5.7 km/l (12–16 mpg)

CONWAY TWITTY
The crooner of such tunes as *It's Only Make Believe*
owned the '64 Riv on these pages, and he personalized
it with his own number plate.

CHUNKY PILLAR
*Hefty rear pillars made
for tricky blind-spots.*

ELECTRONIC BOOT
*One optional extra was a
remote-controlled boot lid,
which was pretty neat for '64.*

TYRES
*Optional whitewalls and
Formula Five chrome-look steel
wheels made a cute car even cuter.*

CLASSIC RIV FRONT
'63 and '64 Rivs have classic exposed double headlights. For reasons best known to themselves, Buick gave '65 cars headlights that were hidden behind electrically-operated, clam-shell doors.

ENGINE
'64s had a 425cid Wildcat V8 that could be tickled up to 360 horses, courtesy of dual four-barrels. *Car Life* magazine tested a '64 Riv with the Wildcat unit and stomped to 60 mph (96 km/h) in a scintillating 7.7 seconds.

ROVER TRANSFER
Buick sold the tooling for the old 401 to Rover, who used it to great success on their Range Rover.

ENGINE OPTION
'65 saw a Gran Sport option with 360 bhp mill, limited slip diff, and "Giro-Poise" roll control.

INTERIOR
The sumptuous Riv was a full four-seater, with the rear seat divided to look like buckets. The dominant V-shaped centre console mushroomed from between the front seats to blend into the dashboard. The car's interior has a European ambience that was quite uncharacteristic for the period.

GRILLE
The grille was inspired by the Ferrari 250GT.

T-BIRD BEATER
High-rolling price of $4,333 was actually $153 cheaper than Ford's T-Bird.

W-SHAPE
The purposeful W-section front could have come straight out of an Italian styling house. The classy Riviera soon became the American Jaguar.

Buick *Riviera (1971)*

THE '63 RIVIERA HAD BEEN one of Buick's best sellers, but by the late Sixties it was lagging far behind Ford's now-luxurious Thunderbird. Mind you, the Riviera easily outsold its stable-mate, the radical front-wheel drive Toronado, but for '71 Buick upped the stakes by unveiling a new Riviera that was a little bit special. The new model had become almost a caricature of itself, now bigger and brasher than it ever was before. Handsome and dramatic, the "boat-tail", as it was nicknamed, had its stylistic roots in the split rear-screen Sting Ray of '63. It was as elegant as Jackie Onassis and as hard-hitting as Muhammad Ali. Its base price was $5,251, undercutting the arch-rival T-Bird by a wide margin.

Designer Bill Mitchell nominated it as his favourite car of all time and, while sales of Rivieras hardly went crazy, at last Buick had a flagship model that was the envy of the industry. It was the coupé in which to make a truly stunning entrance.

ENGINE
The Riviera came with GM's biggest mill, the mighty 455. The even hotter Gran Sport option made the massive V8 even smoother and quieter and offered big-buck buyers a stonking 330 bhp. One reviewer said of the GS-engined car, "there's nothing better made on these shores".

GRILLE
The lines of the boat-tail were not only beautiful at the rear but were carried right through to the thrusting, pointed grille.

LAND OF ENCHANTM
AMY ⬥ 5
72 NEW MEXICO US

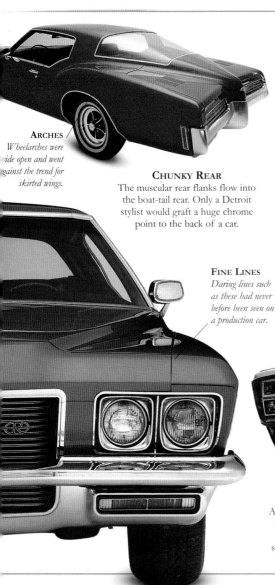

ARCHES
Wheelarches were wide open and went against the trend for skirted wings.

CHUNKY REAR
The muscular rear flanks flow into the boat-tail rear. Only a Detroit stylist would graft a huge chrome point to the back of a car.

FINE LINES
Daring lines such as these had never before been seen on a production car.

SPECIFICATIONS

MODEL Buick Riviera (1971)

PRODUCTION 33,810 (1971)

BODY STYLE Two-door coupé.

CONSTRUCTION Steel body and box-section chassis.

ENGINE 455cid V8.

POWER OUTPUT 315–330 bhp.

TRANSMISSION Three-speed Turbo Hydra-Matic automatic.

SUSPENSION *Front:* independent coil springs; *Rear:* self-levelling pneumatic bellows over shocks.

BRAKES Front discs, rear drums.

MAXIMUM SPEED 201 km/h (125 mph)

0–60 MPH (0–96 KM/H) 8.4 sec

A.F.C. 4.2–5.3 km/l (12–15 mpg)

CABIN
The Seventies cabin was plush but plasticky.

INTERIOR
After 1972, the rear seat could be split 60/40 – pretty neat for a coupé. The options list was infinite and you could swell the car's sticker price by a small fortune. Tilt steering wheel came as standard.

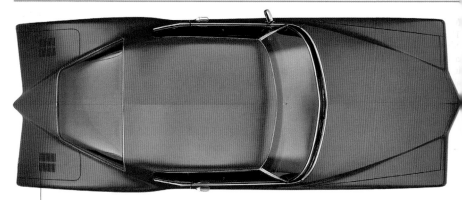

AIR VENTS
Vents were part of the air-conditioning system and unique to '71 Rivieras.

OVERHEAD BEAUTY
The Riviera's styling may have been excessive, but it still made a capacious five-seater, despite the fastback roof line and massive rear window. The 3.1 m (122 in) wheelbase made the '71 boat-tail longer than previous Rivieras.

PILLARLESS STYLE
With the side windows down, the Riv was pillarless, further gracing those swooping lines.

SUPREME STOPPING POWER
The Riviera drew praise for its braking, helped by a Max Trac anti-skid option. The Riv could stop from 96 km/h (60 mph) in 41 m (135 ft), a whole 12 m (40 ft) shorter than its rivals.

TINTED SCREEN
Soft-Ray tinted glass helped keep things cool.

BRAKES
Discs on the front helped create a quality braking system.

MITCHELL TRADEMARK
The rear was a Bill Mitchell
"classic" that had his
trademark stamped all over
it, the GM supremo having
also designed the rear of
the '63 Sting Ray coupé.

REAR VIEW
*View from
rear-view mirror
was slightly
restricted.*

SEATS
*Seating could be
all-vinyl bench
seats with
custom trim or
front buckets.*

REAR SCREEN
*One-piece rear windscreen
curves downwards.*

BOOT RELEASE
*Electric boot releases are not a
modern phenomenon – they were
on the '71 Riviera's options list.*

CADILLAC *Series 62*

WE OWE A LOT TO THE '49 Cadillac. It brought us tail fins and a high-compression V8. Harley Earl came up with those trendsetting rear rudders, and John F. Gordon the performance motor. Between them they created the basic grammar of the post-war American car. In 1949 the one millionth Cad rolled off the production line, and the stunning Series 62 was born. Handsome and quick, it was a complete revelation. Even the haughty British and Italians nodded sagely in admiration and, at a whisker under $3,000, it knocked the competition dead in their tracks. As Cadillac adverts boasted: "The new Cadillac is not only the world's most beautiful and distinguished motor car, but its performance is a challenge to the imagination." The American Dream and the finest era in American cars began with the '49 Cadillac.

INTERIOR
The cabin was heavily chromed, and oozed quality. Colours were grey-blue or brown with wool carpets to match, and leather or cloth seats. Steering was Saginaw, with standard four-speed auto transmission.

UNDER THE BONNET
While styling was similar to that of the '48 model, the new OHV V8 in the '49 was an innovation.

CADDY INSPIRATION

1948 was the year of the fin and the year of the crème des Cads. Cadillac designers Bill Mitchell, Harley Earl, Frank Hershey, and Art Ross had been smitten by a secret P-38 Lockheed Lightning fighter plane. Cadillac also had Ed Cole's OHV V8, some 10 years in the making. With a brief to reduce weight and increase compression, the end result was an engine with more torque and better mileage than any other at the time.

WINDSCREEN
Curved windscreen was a novelty for a 1949 car.

SECRET CAP
Fuel filler-cap was hidden under tail light, a Cadillac trait since 1941.

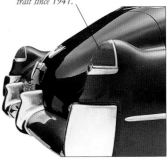

CADDY BADGING
The "V" emblem below the crest denoted V8 power; the basic badge design remained unaltered until 1952.

TAIL VIEW
The plane-inspired rear fins became a Caddy trademark and would reach a titanic height on '59 models.

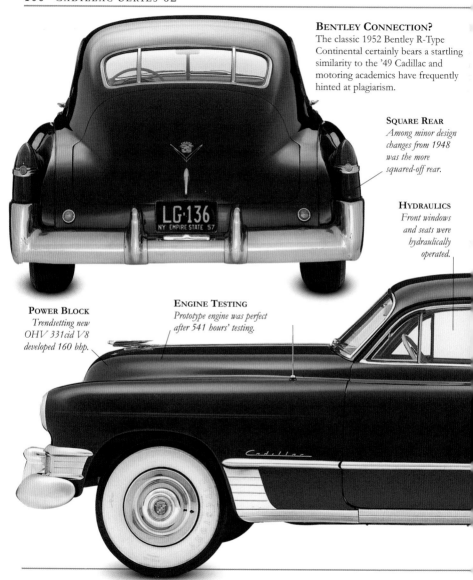

BENTLEY CONNECTION?
The classic 1952 Bentley R-Type Continental certainly bears a startling similarity to the '49 Cadillac and motoring academics have frequently hinted at plagiarism.

SQUARE REAR
Among minor design changes from 1948 was the more squared-off rear.

HYDRAULICS
Front windows and seats were hydraulically operated.

POWER BLOCK
Trendsetting new OHV 331cid V8 developed 160 bhp.

ENGINE TESTING
Prototype engine was perfect after 541 hours' testing.

LG·136
NY EMPIRE STATE 57

BELIEVE THE HYPE
Cadillac advertisements trumpeted that the '49 was "the world's most beautiful car", and the simple yet elegant styling caught the public's imagination.

FINE LINES
Glorious tapering roof line.

GRILLE
Grille was heavier on the '49 than on the '48.

TYRES
Tyres ran at only 24 psi, making unassisted steering heavy for the driver.

DECORATION
Chrome slashes were inspired by aircraft air intakes.

CLASSIC STYLING
Hugely influential body design was penned by Harley Earl and Julio Andrade at GM's styling studios. Many of the '49 features soon found themselves on other GM products such as Oldsmobile and Buick.

CADILLAC *Eldorado Convertible (1953)*

FOR 1950s AMERICA, CARS DID NOT come much more glamorous than the 1953 Eldorado. "A car apart – even from other Cadillacs", assured the advertising copy. The first Caddy to bear the Eldo badge, it was seen as the ultimate and most desirable American luxury car, good enough even for Marilyn Monroe and Dwight Eisenhower Conceived as a limited edition, the '53 brought avant-garde styling cues from Harley Earl's Motorama Exhibitions. Earl was Cadillac's inspired chief designer, while Motoramas were yearly futuristic car shows where his whims of steel took on form. At a hefty $7,750, nearly twice as much as the regular Cadillac Convertible and five times as much as an ordinary Chevrolet, the '53 was special. In 1954, Cadillac cut the price by 50 per cent and soon Eldorados were leaving showrooms like heat-seeking missiles. Today collectors regard the '53 as the one that started it all – the first and most fabulous of the Eldorados.

AIR-CON WEIGHT
Air-conditioning boosted the car's weight to 2,177 kg (4,800 lb), but top speed was still a brisk 187 km/h (116 mph).

POWER TOPPERS
At the time the '53 was America's most powerful car, with a cast-iron V8, four-barrel carburettor, and wedge cylinder head. With the standard convertible weighing 136 kg (300 lb) less, the Eldorado was actually the slowest of the Cadillacs.

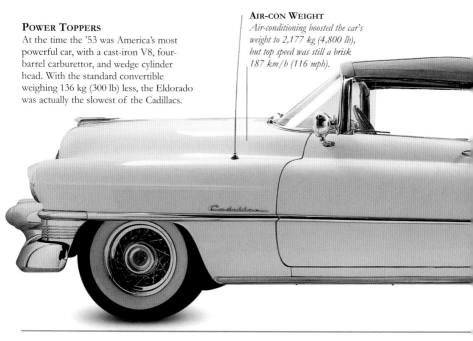

SPECIFICATIONS

MODEL Cadillac Eldorado Convertible (1953)

PRODUCTION 532 (1953)

BODY STYLE Five-seater convertible.

CONSTRUCTION Steel bodywork.

ENGINE 5424cc V8.

POWER OUTPUT 210 bhp at 4150 rpm.

TRANSMISSION Three-speed Hydra-Matic Dual-Range automatic.

SUSPENSION *Front:* independent MacPherson strut; *Rear:* live axle with leaf springs.

BRAKES Front and rear drums.

MAXIMUM SPEED 187 km/h (116 mph)

0–60 MPH (0–96 KM/H) 12.8 sec

0–100 MPH (0–161 KM/H) 20 sec

A.F.C. 5–7 km/l (14–20 mpg)

FUTURISTIC STYLING

The twin exhausts emerge from the rear bumper – the beginnings of "jet-age" styling themes which would culminate in the outrageous 107-cm (42-in) fins on the 1959 Cadillac Convertible *(see pages 106–09).*

MATERIAL
Hood was made of Orlon acrylic.

SLICK DESIGN
The hood disappeared neatly below a steel tonneau panel, giving the Eldorado a much cleaner uninterrupted line than other convertibles.

SPARE WHEEL
The boot-mounted spare wheel was an after-market continental touring kit.

TYRES
Swish whitewall tyres and chrome wire wheels were standard on the Eldorado Convertible.

TOP OF THE RANGE
As Cadillac's finest flagship, the Eldorado had image by the bucketful. The 331 cubic inch V8 engine was the most powerful yet, and the body line was ultra sleek.

WINDSCREEN
The standard Cadillac wrap-around windscreen was first seen on the '53.

TWO-WAY MIRROR
The heavily chromed, hand-operated swivelling spotlight doubled up as a door mirror.

CHROME STYLING
Dagmars were so-called after a lushly upholstered starlet of the day.

CAL
DRM

AERIAL
Aerial picked up reception for self-tuning radio.

DASHBOARD
Standard equipment on the Eldo Convertible was Hydra-Matic transmission, hydraulic window lifts, leather and cloth upholstery, tinted glass, vanity and side mirrors, plus a self-tuning radio.

BODY COLOUR
Colours available were Alpine White, Aztec Red, Azure Blue, and Artisan Ochre.

CADILLAC *Convertible*

NO CAR BETTER SUMS UP AMERICA at its peak than the 1959 Cadillac – a rocket-styled starship for orbiting the galaxy of new freeways in the richest and most powerful country on earth. With 107 cm (42 in) fins, the '59 Cad marks the zenith of American car design. Two tonnes in weight, 6.1 m (20 ft) long, and 1.83 m (6 ft) wide it oozed money, self-confidence, and unchallenged power. Under a bonnet almost the size of Texas nestled an engine almost as big as California. But while it might have looked like it was jet-powered, the '59 handled like the *Amoco Cadiz*. No matter. The '59 Cad will always be remembered as a glorious monument to the final years of shameless American optimism. And for a brief, hysterical moment the '59 was the pre-eminent American motor car, the ultimate in crazed consumerism. Not a car, but an exemplar of its time that says more about Fifties America than a trunk of history books. The '59 *was* the American Dream.

HALLOWED STATUS
With tail fins that rose a full 1.07 m (3½ ft) off the ground, the '59 is an artefact, a talisman of its times. Not a car, but a styling icon, wonderfully representative of the end of an era – the last years of American world supremacy and an obsession with space travel and men from Mars.

WINDSCREEN
Steep, wrap-around windscreen could have come straight out of a fighter plane.

QUARTERLIGHTS
Chrome door quarterlights could be swivelled from inside the car.

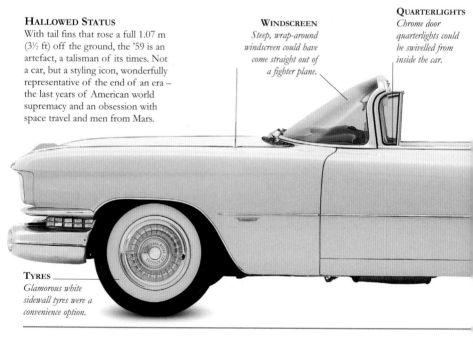

TYRES
Glamorous white sidewall tyres were a convenience option.

HOOD
With hood furled, the Cad had an uninterrupted, dart-like profile.

LIGHTS
Egg-shaped ruby tail lights are pure jet age.

EXCESS REAR
Commentators at the time actually thought the '59 too garish. So did Cadillac, who took 15.5 cm (6 in) off the fins in the following model year.

DOORS
Massive slab-sided doors gave an easy entrance and exit.

BOOT
Boot was cavernous and could hold five wheels.

XSU·385 PENNSYLVANIA

EXTRAVAGANT LENGTH
The '59's length meant that its turning circle was a massive 7.3 m (24 ft).

SPACIOUS
Interior was vast, a true six-seater with acres of room.

BONNET STATUS

With a bonnet the size of an aircraft carrier, the '59 Cad was perfect for a society where a car's importance was defined by the length of its nose. The price to pay for such excess was that the front end was notorious for vibration. To help with the comfort factor, electrically operated seats, windows, and boot could all be ordered.

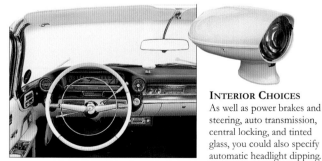

INTERIOR CHOICES

As well as power brakes and steering, auto transmission, central locking, and tinted glass, you could also specify automatic headlight dipping.

ENGINE

The monster 6.3-litre V8 engine had a cast-iron block, five main bearings, and hydraulic valve lifters, pushing out a not inconsiderable 325 bhp at 4800 rpm.

SPECIFICATIONS

MODEL Cadillac Eldorado Convertible (1959)

PRODUCTION 11,130 (1959)

BODY STYLE Two-door, six-seater convertible.

CONSTRUCTION X-frame chassis, steel body.

ENGINE 6.3-litre (390cid) V8.

POWER OUTPUT 325/345 bhp at 4800 rpm.

TRANSMISSION GM Hydra-Matic three-speed automatic.

SUSPENSION All-round coil springs with optional Freon-12 gas suspension.

BRAKES Four-wheel hydraulic power-assisted drums.

MAXIMUM SPEED 180 km/h (112 mph)

0–60 MPH (0–96 KM/H) 10.3 sec

0–100 MPH (0–161 KM/H) 23.1 sec

A.F.C. 2.8 km/l (8 mpg)

HIDDEN LIGHTS

Extravagant mounds of chrome might look like turbines but conceal reversing lights.

TAIL VIEW

The '59's outrageous fins, which are the highest of any car in the world, are accentuated by its very low profile – 8 cm (3 in) lower than the '58 model's already modest elevation.

CADILLAC *Eldorado* (1976)

BY 1976, CADILLACS HAD BECOME so swollen that they ploughed through corners, averaged 4.6 km/l (13 mpg), and were as quick off the line as an M24 tank. Despite a massive 500cid V8, output of the '76 Eldo was a lowly 190 brake horsepower, with a glacial top speed of just 175 km/h (109 mph). Something had to change and Cadillac's response had been the '75 Seville. But the '76 Eldo marked the end of an era for another reason – it was the last American convertible. Cadillac were the final automobile manufacturer to delete the rag-top from their model line-up and, when they made the announcement that the convertible was to be phased out at the end of '76, the market fought to buy up the last 200. People even tried to jump the queue by claiming they were distantly related to Cadillac's founder. One 72-year-old man in Nebraska bought six. A grand American institution had quietly passed away.

TRADITIONAL SET-UP
Big and slab-sided, the '76 Eldo used a front-wheel drive arrangement that had first been used on the '67 Eldorado and is still used today. The '76 convertible had big vital statistics, measuring 5.7 m (225 in) long, 2 m (80 in) wide, and costing $10,354.

FITTINGS
Interiors could be specified in Merlin Plaid, lush velour, Mansion Knit, or 11 types of Sierra Grain leather.

FUNKY MIRROR
The heavy chrome adjustable door mirror was electrically operated and incorporated a thermometer that displayed the outside temperature.

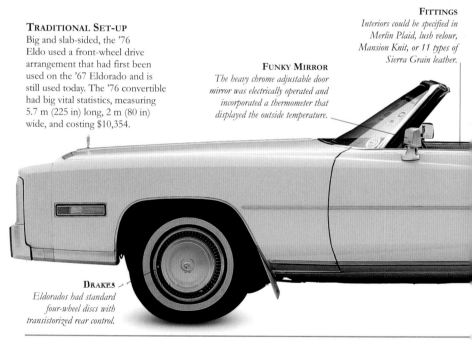

BRAKES
Eldorados had standard four-wheel discs with transistorized rear control.

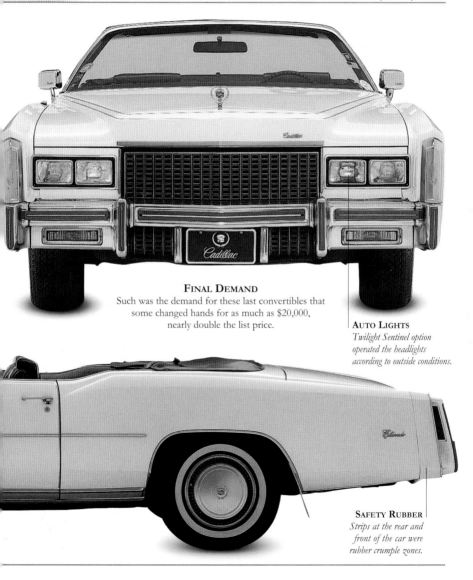

FINAL DEMAND
Such was the demand for these last convertibles that
some changed hands for as much as $20,000,
nearly double the list price.

AUTO LIGHTS
*Twilight Sentinel option
operated the headlights
according to outside conditions.*

SAFETY RUBBER
*Strips at the rear and
front of the car were
rubber crumple zones.*

ECONOMY CLASS

Raised compression ratios and a recalibrated carburettor gave the Eldo better fuel economy than might be expected from such a mammoth block. Hydro-Boost power brakes were needed to stop the 2,337 kg (5,153 lb) colossus.

SUSPENSION
Independent coil springs were complemented by automatic level control.

WOOD
Interior wood was called "distressed pecan grain".

INTERIOR

Technically advanced options were always Cadillac's forte. The Eldo was available with an airbag, Dual Comfort front seats with fold-down armrests, and a six-way power seat.

COLOUR CHOICE
Eldos could be ordered in 21 body colours.

SPECIFICATIONS

MODEL Cadillac Eldorado Convertible (1976)

PRODUCTION 14,000 (1976)

BODY STYLE Two-door, six-seater convertible.

CONSTRUCTION Steel body and chassis.

ENGINE 500cid V8.

POWER OUTPUT 190 bhp.

TRANSMISSION Three-speed Hydra-Matic Turbo automatic.

SUSPENSION Front and rear independent coil springs with automatic level control.

BRAKES Four-wheel discs.

MAXIMUM SPEED 175 km/h (109 mph)

0–60 MPH (0–96 KM/H) 15.1 sec

A.F.C. 4.6 km/l (13 mpg)

ENGINE
lready strangled by
mission pipery, the need to
maximize every gallon meant that the
big 500bhp V8 was embarrassingly lethargic
when it came to speed. Even lower ratio
rear axles were used to boost mileage.

SPACE
*Even with the hood up, the
Eldo was gargantuan inside.*

CONVERTOR
*All Eldorados
had a catalytic
convertor as
standard.*

CADILLAC NAME
The Cadillac shield harks
back to 1650 and the original
French Cadillac family.
French model names were
used in 1966 with the Calais
and DeVille ranges.

REFLECTORS
*Slightly superfluous in that
not many drivers would
miss this giant on the road.*

CHEVROLET *Corvette (1954)*

A CARICATURE OF A EUROPEAN roadster, the first Corvette of 1953 was more show than go. With typical arrogance, Harley Earl was more interested in the way it looked than the way it went. But he did identify that car consumers were growing restless and saw a huge market for a new type of auto opium. With everybody's dreams looking exactly the same, the plastic 'Vette brought a badly needed shot of designed-in diversity. Early models may have been cramped and slow, but they looked like they'd been lifted straight off a Motorama turntable, which they had. Building them was a nightmare though, and for a while GM lost money on each one. Still, nobody minded because Chevrolet now had a new image – as the company that came up with the first American sports car.

EXHIBITION SUCCESS

The 'Vette's shape was based on the 1952 EX-122 show car, and this was one of the few Motorama dream cars to go into production virtually unchanged. The original plan to produce the 'Vette in steel was shelved after widespread acclaim for the glass-fibre body from visitors to Motorama.

PERFORMANCE

Performance was not in the Jaguar XK120 league, with a modest 172 km/h (107 mph) top speed.

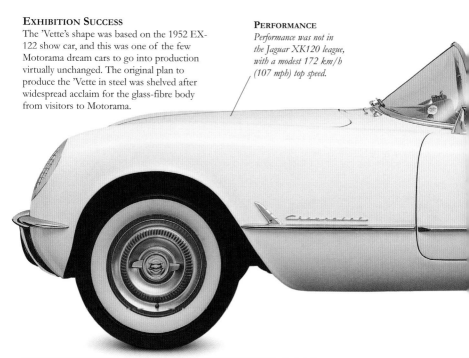

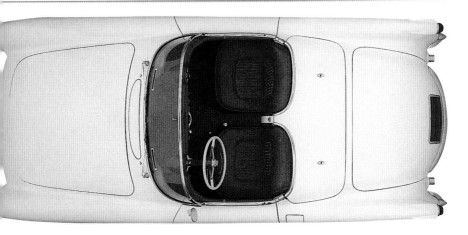

OVERVIEW

The cleverly packaged glass-fibre body was rather tricky to make, with no less than 46 different sections. The soft-top folded out of sight below a neat lift-up panel.

INTERNAL HANDLES

Like the British sports cars it aped, the '54 'Vette's door handles lived on the inside.

SUSPENSION

Outboard-mounted rear leaf springs helped cornering stability.

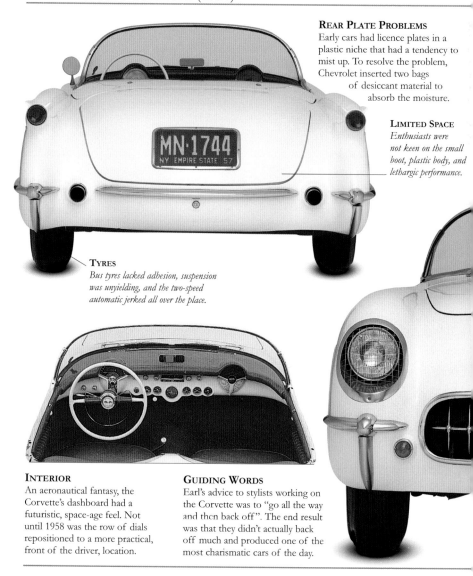

REAR PLATE PROBLEMS
Early cars had licence plates in a plastic niche that had a tendency to mist up. To resolve the problem, Chevrolet inserted two bags of desiccant material to absorb the moisture.

LIMITED SPACE
Enthusiasts were not keen on the small boot, plastic body, and lethargic performance.

TYRES
Bus tyres lacked adhesion, suspension was unyielding, and the two-speed automatic jerked all over the place.

INTERIOR
An aeronautical fantasy, the Corvette's dashboard had a futuristic, space-age feel. Not until 1958 was the row of dials repositioned to a more practical, front of the driver, location.

GUIDING WORDS
Earl's advice to stylists working on the Corvette was to "go all the way and then back off". The end result was that they didn't actually back off much and produced one of the most charismatic cars of the day.

ENGINE

The souped-up Blue Flame Six block may have had triple carburettors, higher compression, and a high-lift cam, but it was still old and wheezy. 'Vettes had to wait until 1955 for the V8 they deserved.

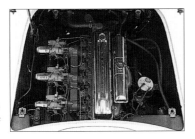

SPECIFICATIONS

MODEL Chevrolet Corvette (1954)

PRODUCTION 3,640 (1954)

BODY STYLE Two-door, two-seater sports.

CONSTRUCTION Glass-fibre body, steel chassis.

ENGINE 235.5cid straight-six.

POWER OUTPUT 150 bhp.

TRANSMISSION Two-speed Powerglide automatic.

SUSPENSION *Front:* coil springs; *Rear:* leaf springs with live axle.

BRAKES Front and rear drums.

MAXIMUM SPEED 172 km/h (107 mph)

0–60 MPH (0–96 KM/H) 8–12 sec

A.F.C. 7 km/l (20 mpg)

BODY COLOUR
Oddly enough, 80 per cent of all '54 Corvettes were painted white.

GUARDS
Stone-guards on lights were culled from European racing cars, but criticized for being too feminine.

ITALIAN SMILE
Earl admitted that the shark-tooth grille was robbed from contemporary Ferraris.

BUMPERS
Impact protection may have been vestigial, but the glass-fibre body took knocks well.

MN·1744
NY EMPIRE STATE 57

CHEVROLET *Bel Air (1957)*

CHEVROLET CALLED THEIR '57 LINE "sweet, smooth, and sassy", and the Bel Air was exactly what America wanted – a junior Cadillac. Finny, trim, and handsome, and with Ed Cole's Super Turbo-Fire V8, it boasted one of the first production engines to pump out one horsepower per cubic inch, and was the first mass-market "fuelie" sedan with Ramjet injection. Chevy copywriters screamed "the Hot One's even hotter", and Bel Airs became kings of the street. Production that year broke the 1½ million barrier and gave Ford the fright of their life. The trouble was that the "Hot One" was forced to cool it when the Automobile Manufacturers' Association urged car makers to put an end to their performance hysteria. Today, the Bel Air is one of the most widely coveted US collector's cars and the perfect embodiment of young mid-Fifties America. In the words of the Billie Jo Spears song, "Wish we still had her today; the good love we're living, we owe it to that '57 Chevrolet".

BODY STYLE
Other body styles available included a two-door hardtop.

POPULAR AND STYLISH
At $2,511, the Bel Air Convertible was the epitome of budget-priced good taste, finding 47,562 eager buyers. Low, sleek, and flashy, it could almost out-glam the contemporary Caddy rag-top.

BUICK STYLE
The Bel Air's Ventiports only lasted a couple of years.

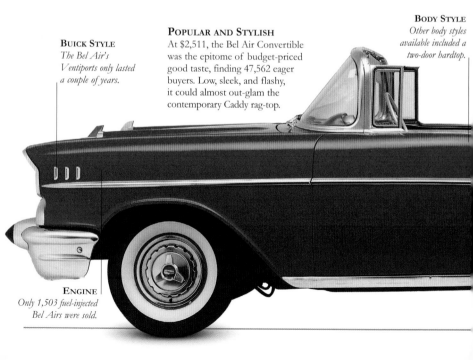

ENGINE
Only 1,503 fuel-injected Bel Airs were sold.

ORNAMENTATION
The rather clumsy bomb-sight bonnet ornament could be fairly described as the '57 Bel Air's only minor stylistic blemish. The public liked it, though.

FRENCH DECORATION
Chevrolet's fleur-de-lis, a reminder of their French roots.

SAFETY MEASURES
Seat belts and shoulder harnesses were available on the lengthy options list.

PERFECTLY FORMED
Immediately after it was introduced, it was rightly hailed as a design classic. Elegant, sophisticated, and perfectly proportioned, the '57 Bel Air is one of the finest post-war American autos of all.

LONGER MODEL
The '57 Bel Air was 6.3 cm (2½ in) longer than the '56 model.

A TRUE CLASSIC

The '57 Bel Air sums up America's most prosperous decade better than any other car of the time. Along with hula-hoops, drive-in movies, and rock 'n' roll, it has become a Fifties icon. It was loved then because it was stylish, solid, sporty, and affordable, and it's loved now for more or less the same reasons; plus it simply drips with nostalgia.

INTERIOR
The distinctive two-tone interiors were a delight. Buyers could opt for a custom colour interior, power convertible top, tinted glass, vanity mirror, ventilated seat pads, power windows, and even a tissue dispenser.

POWER OPTION
The Bel Air Convertible could be fitted with an optional power-operated top.

SPEEDOMETER
Speedo read to 120, and larger-engined models nearly broke through the dial.

BEHIND THE WHEEL
The small-block Turbo-Fire V8 packed 185 bhp in base two-barrel trim and 270 bhp with the optional Rochester four-barrel. Ramjet injection added a hefty $500 to the sticker price.

AIR STYLE
Chevrolet, like every other US motor manufacturer at the time, were keen to cash in on the jet age, but in reality this '55 Bel Air four-door sedan looks positively dumpy next to the fighter plane.

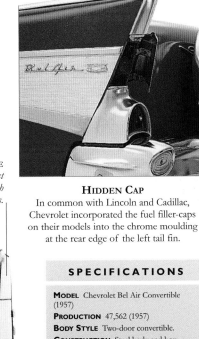

HIDDEN CAP
In common with Lincoln and Cadillac, Chevrolet incorporated the fuel filler-caps on their models into the chrome moulding at the rear edge of the left tail fin.

RESTRAINED FINNAGE
Subtle rear fins are almost demure compared with other contemporary efforts.

SPECIFICATIONS

MODEL Chevrolet Bel Air Convertible (1957)

PRODUCTION 47,562 (1957)

BODY STYLE Two-door convertible.

CONSTRUCTION Steel body and box-section chassis.

ENGINES 265cid, 283cid V8s.

POWER OUTPUT 162–283 bhp (283cid V8 fuel injected).

TRANSMISSION Three-speed manual with optional overdrive, optional two-speed Powerglide automatic, and Turboglide.

SUSPENSION *Front:* independent coil springs; *Rear:* leaf springs with live axle.

BRAKES Front and rear drums.

MAXIMUM SPEED 145–193 km/h (90–120 mph)

0–60 MPH (0–96 KM/H) 8–12 sec

A.F.C. 5 km/l (14 mpg)

CHEVROLET *Bel Air Nomad (1957)*

IF YOU THOUGHT BMW AND MERCEDES were first with the sporting uptown carry-all, think again. Chevrolet kicked off the genre as far back as 1955. The Bel Air Nomad was a development of Harley Earl's dream-car wagon based on the Chevrolet Corvette and although it looked like other '55 Bel Airs, the V8 Nomad was the most expensive Chevy ever. But despite the fact that *Motor Trend* described the '57 Nomad as "one of the year's most beautiful cars", with only two doors its appeal was limited, its large glass area made the cabin too hot, and the twinkly tailgate let in water. No surprise then that it was one of Chevy's least popular models. Sales never broke the magic 10,000 barrier and, by 1958, the world's first sportwagon, and now a milestone car, had been dropped.

ROOF FIRST
The Nomad was the first car to use non-structural corrugations on the roof.

INTERIOR
Two-tone trim could be complemented by power seat, tinted glass, and seat belts.

STYLE REVIVAL
The Nomad was essentially a revival of the original Town and Country theme and a reaction against the utilitarian functionalism of the boxy wooden wagons that had become ubiquitous in suburban America.

ENGINE
Base unit was a 235cid six; grunty 265cid V8 was available.

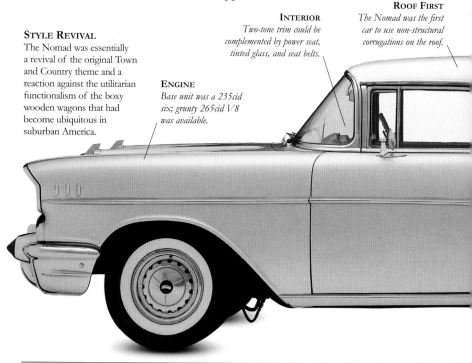

SPECIFICATIONS

MODEL Chevrolet Bel Air Nomad (1957)

PRODUCTION 6,103 (1957)

BODY STYLE Two-door station wagon.

CONSTRUCTION Steel body and chassis.

ENGINES 235cid six, 265cid V8.

POWER OUTPUT 123–283 bhp.

TRANSMISSION Three-speed manual with overdrive, two-speed Powerglide automatic, and optional Turboglide.

SUSPENSION *Front:* coil springs; *Rear:* leaf springs.

BRAKES Front and rear drums.

MAXIMUM SPEED 145–193 km/h (90–120 mph)

0–60 MPH (0–96 KM/H) 8–11 sec

A.F.C. 5.3–6.7 km/l (15–19 mpg)

IMMEDIATE HIT

Unveiled in January 1954, the Motorama Nomad – created by Chevy stylist Carl Renner – was such a hit that a production version made it into the '55 brochures.

DECORATED TAIL

The classic Harley Earl embellished tailgate was lifted straight from the Motorama Corvette and was widely praised.

'VETTE LINES

Motorama 'Vette roof line was adapted for production Nomads in just two days.

CHEVROLET *3100 Stepside*

CHEVY WERE ON A HIGH in the mid-Fifties. With the 'Vette, the Bel Air, and their new V8, they were America's undisputed top car manufacturer. A boundless optimism percolated through all divisions, even touching such prosaic offerings as trucks. And the definitive Chevy carry-all has to be the '57 pick-up. It had not only that four-stroke overhead-valve V8 mill, but also various options and a smart new restyle. Small wonder it was nicknamed "a Cadillac in drag". Among the most enduring of all American design statements, the '57 had clean, well-proportioned lines, a minimum of chrome, and integrated wings. Chevrolet turned the pick-up from a beast of burden into a personalized workhorse complete with all the appurtenances of gracious living usually seen in a boulevard cruiser.

ENGINE
The small-block V8 produced 150 bhp and could cruise at 113 km/h (70 mph). From '55, all Chevys used open-drive instead of an enclosed torque-tube driveline.

WRAP-AROUND SCREEN
De Luxe models had a larger, wrap-around windscreen, and two-tone seats, door trims, and steering wheel.

INTERIOR
The Stepside was as stylized inside as out, with a glovebox, heavy chrome switches, and a V-shaped speedo.

TIMBER BED

Wooden-bed floors helped to protect
the load area and added a quality feel
to Chevy's Stepside.

SPECIFICATIONS

MODEL Chevrolet 3100 Stepside (1957)

PRODUCTION Not available.

BODY STYLE Two-seater, short-bed
pick-up.

CONSTRUCTION Steel body and chassis.

ENGINES 235cid six, 265cid V8.

POWER OUTPUT 130–145 bhp.

TRANSMISSION Three-speed manual
with optional overdrive, optional
three-speed automatic.

SUSPENSION *Front:* coil springs;
Rear: leaf springs.

BRAKES Front and rear drums.

MAXIMUM SPEED 129 km/h
(80 mph)

0–60 MPH (0–96 KM/H) 17.3 sec

A.F.C. 6 km/l (17 mpg)

MULTIPLE CHOICES

Chevy's '57 pick-ups can be
identified by the new trapezoid grille
and a flatter bonnet than '56 models.
Buyers had a choice of short or long
pick-up, De Luxe or standard trim,
and 11 exterior colours. Engines
were the 235cid Thriftmaster six
or the 265cid Trademaster V8.

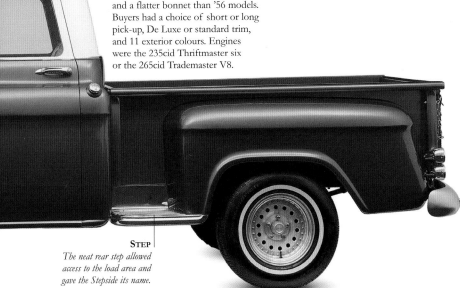

STEP

*The neat rear step allowed
access to the load area and
gave the Stepside its name.*

CHEVROLET *Impala*

IN THE SIXTIES, unbridled consumerism began to wane. America turned away from the politics of prosperity and, in deference, Chevrolet toned down its finny Impala. The '59's gothic cantilevered batwings went, replaced by a much blunter rear deck. WASP America was developing a social conscience and Fifties excess just wasn't cool anymore. Mind you, the '60 Impala was no shrinking violet. Tired of gorging on gratuitous ornamentation, US motorists were offered a new theology – performance. Freeways were one long concrete loop, premium gas was cheap, and safety and environmentalism were a nightmare still to come. For $333, the Sports Coupe could boast a 348cid, 335 bhp Special Super Turbo-Thrust V8. The '59 Impala was riotous and the '60 stylistically muddled, but within a year the unruliness would disappear altogether. These cross-over Chevrolets are landmark cars – they ushered in a new decade that would change America and Americans forever.

RESTRAINED STYLING
The front of the Impala was meant to be quiet and calm and a million miles from the deranged dentistry of mid-Fifties grille treatments. The jet-fighter cockpit and quarter-panel missile ornaments were eerie portents of the coming decade of military intervention.

LUXURY EXTRAS
Chevy's trump card was an option list normally found on luxury autos, like air-conditioning, power steering and windows, and six-way power seat.

NATION'S FAVOURITE
The Impala was America's best-selling model in 1960.

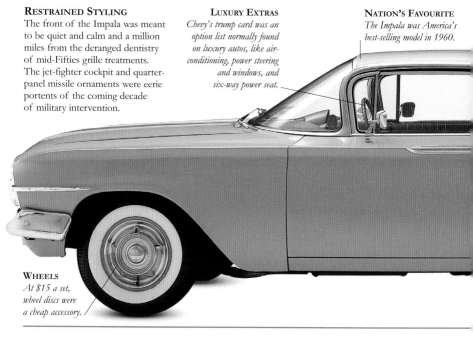

WHEELS
At $15 a set, wheel discs were a cheap accessory.

SPACE STYLE
Chevy's ad men sold the '60 Impala on "Space-Spirit-Splendor".

CLASSY REAR END
Triple tail lights and a vertically ribbed aluminium rear beauty panel helped to sober up the Impala's rear end. It was still a class act and a lot glitzier than the Bel Air's plainer tail.

QUALITY RIDE
The Impala's coil spring suspension was superior to the leaf-spring rear system found on rival cars.

IMMEDIATE SUCCESS
The Impala debuted in '58 as a limited edition but went on to become the most popular car in '60s America.

COUPE BEAUTY
The Sport Coupe is the prettiest Impala, with appealing proportions and a sleek dart-like symmetry.

EXHAUST OPTION
Dumping all that lead were twin exhausts, a bargain $19 option.

STEERING WHEEL
The sporty steering wheel was inspired by the Corvette.

INTERIOR
Inside, the Impala was loaded with performance metaphor: central speedo, four gauges, and a mock sports steering wheel with crossed flags. This car incorporates power windows and dual Polaroid sun visors.

TRIPLE LIGHTS
The triple tail lights had disappeared in '59 but returned for the '60 model; they went on to become a classic Impala styling cue.

LENGTHY FRAME
Impalas were big, riding on a 302-cm (119-in) wheelbase.

TYRES
Slick whitewalls were yours for just $36.

TAME FINS
The '60 Impala sported much tamer Spread Wing fins that aped a seagull in flight. They were an answer to charges that the '59's uproarious rear end was downright dangerous.

MODEL RANGE

Body styles were four-door sports sedan, pillarless sport coupe, stock four-door sedan, and convertible.

EXTRA BOOST

Impalas could be warmed up considerably with some very special engines.

RACING MODELS

The Impala impressed on circuits all over the world. In 1961, some models were deemed hot enough to dice with European track stars like the Jaguar Mark II, as driven by Graham Hill.

ENGINE

Two V8 engine options offered consumers seven heady levels of power, from 170 to 335 horses. Cheapskates could still specify the ancient Blue Flame Six, which wheezed out a miserly 135 bhp. Seen here is the 185 bhp, 283cid V8. Impalas could be invigorated with optional Positraction, heavy-duty springs, and power brakes.

CHEVROLET *Corvette Sting Ray (1966)*

THE CHEVROLET CORVETTE IS AMERICA'S native sports car. The "plastic fantastic", born in 1953, is still fantastic more than half-a-decade later. Along the way, in 1992, it notched up a million sales, and it is still hanging in there. Admittedly it has mutated over the years, but it has stayed true to its roots in one very important aspect. Other American sports car contenders, like the Ford Thunderbird *(see pages 274–77),* soon abandoned any sporting pretensions, adding weight and middle-aged girth, but not the Corvette. All Corvette fanciers have their favourite eras: for some it is the purity of the very first generation from 1953; others favour the glamorous 1956–62 models; but for many the Corvette came of age in 1963 with the birth of the Sting Ray.

HIDDEN LIGHTS
Twin, pop-up headlights were hidden behind electrically operated covers; more than a gimmick, they aided aerodynamic efficiency.

BADGING
Corvettes from 1963 to 1967 were known as Sting Rays; the restyled 1968 model *(see pages 142–45)* was re-badged as Stingray, one word. The chequered flag on the front of the bonnet denotes sporting lineage, while the red flag bears the GM logo and a fleur-de-lis.

CHASSIS
New chassis frame was introduced in 1963.

INTERIOR

The Batmobile-style interior, with twin-hooped dash, is carried over from earlier Corvettes but updated in the Sting Ray. The deep-dished, wood-effect wheel comes close to the chest and power steering was an option.

SEATING
Seats were low and flat, rather than figure-hugging.

SPECIFICATIONS

MODEL Chevrolet Corvette Sting Ray (1963–69)

PRODUCTION 118,964

BODY STYLES Two-door sports convertible or fastback coupé.

CONSTRUCTION Glass-fibre body; X-braced pressed-steel box-section chassis.

ENGINES OHV V8, 5359cc (327cid), 6495cc (396cid), 7008cc (427cid).

POWER OUTPUT 250–375 bhp (5359cc), 390–560 bhp (7008cc).

TRANSMISSION Three-speed manual, opt'l. four-speed manual or Powerglide auto.

SUSPENSION Independent all round. *Front:* Unequal-length wishbones with coil springs; *Rear:* Transverse leaf.

BRAKES Drums to 1965, then discs all round.

MAXIMUM SPEED 245 km/h (152 mph, 7008cc).

0–60 MPH (0–96 KM/H) 5.4 sec (7008cc)

0–100 MPH (0–161 KM/H) 13.1 sec (7008cc)

A.F.C. 3–5.7 km/l (9–16 mpg)

BRAKES
In 1965 the Sting Ray got four-wheeled disc brakes in place of all-round drums.

A MITCHELL CLASSIC

The Sting Ray was a bold design breakthrough, giving concrete expression to many of the ideas of new GM styling chief, Bill Mitchell. He reputedly regarded the 1963 Sting Ray as his finest piece of work. More than half of all production was in convertible roadsters, for which a hardtop was an option.

OVERHEAD VIEW

You can tell this is a "small block" engine – the bonnet power bulge was widened to accommodate the "big block" unit. Three-speed manual transmission was standard, with two-speed automatic and three types of manual four-speed shift optional.

ENGINE OPTIONS

Sting Rays came in three engine sizes – naturally all V8s – with a wide range of power options from 250 bhp to more than twice that. This featured car is a 1966 Sting Ray with "small block" 5359cc V8 and Holley four-barrel carb.

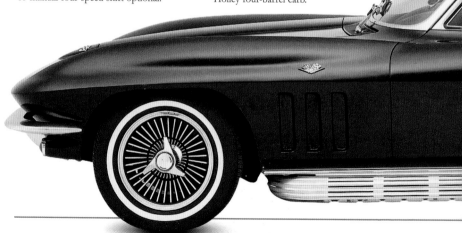

LIMITED BOOT
*Fuel tank and spare
tyre took up most of
the boot space.*

HARDTOP OPTION
Until 1963, all Corvettes were open roadsters, but
with the arrival of the Sting Ray, a fixed-head coupé
was now also available. The distinctive two-piece back
window used on the 1963 model makes it the most
sought-after fixed-head Sting Ray.

SIDE EXHAUST
*Aluminium strip concealed
side-mounted exhaust option.*

CHEVROLET *Corvair Monza*

BY 1960, SALES OF DINOSAURS were down, small-car imports were up, and Detroit finally listened to a market screaming for economy compacts. Then along came Chevrolet's adventurous answer to the Volkswagen Beetle, the pretty, rear-engined Corvair, which sold for half the price of a Ford Thunderbird. But problems soon arose. GM's draconian cost-cutting meant that a crucial $15 suspension stabilizing bar was omitted, and early Corvairs handled like pigs. The suspension was redesigned in '65, but it was too late. Bad news also came in the form of Ralph Nader's book *Unsafe at Any Speed*, which lambasted the Corvair. The new Ford Mustang, which had become the hot compact, didn't help either. By 1969, it was all over for the Corvair. GM's stab at downsizing had been a disaster.

IMPRESSING THE PRESS
After very few styling changes for the first five years, the new body design for '65 had a heavy Italian influence with smooth-flowing, rounded lines that impressed the motoring press. *Car and Driver* magazine called it "the best of established foreign and domestic coachwork".

WING MIRROR
Shatter-resistant wing mirror came as standard.

RAG-TOP NUMBERS
Only 26,000 convertible were sold in '65.

MAINLY AUTOMATIC
Despite the public's interest in economy, 53 per cent of all Corvairs had automatic transmission.

WHEELS
Wire wheel covers were a pricey $59 option.

HOOD
Most tops were manually operated and stowed behind a fabric tonneau, but this model has the $54 power top option.

INTERIOR COLOURS
A choice of eight interior colours included black, fawn, and saddle.

COLOURS
Buyers could choose from 15 exterior colours, a number of which were only available on the Corvair Monza.

REAR-ENGINED
Engine lived here – 95 bhp was dire, 110 fun, and 140 wild. Turbocharged versions could crack 185 km/h (115 mph).

INITIAL SUCCESS
The new longer, wider, and lower Corvair initially sold well but floundered from 1966 in the face of the rival Ford Mustang and Nader's damning book.

REAR SEAT
The rear seat folded down in the Sport Coupe and Sport Sedan models but not in the convertible.

SUSPENSION
The post-'65 Corvair had Corvette-type fully independent rear suspension via upper and lower control arms and coil springs.

EXHAUST
Long-life exhaust system consisted of aluminized silencer.

STORAGE SPACE
Rear-engined format meant that storage space under the bonnet was massive.

INTERIOR
The all-vinyl interior was very European, with bucket seats and telescopic steering column. The restrained steering wheel and deep-set instruments could have come straight out of a BMW. The dials were recessed to reduce glare and deep-twist carpeting added an air of luxury to the cockpit. Options on offer included a windscreen-mounted automatic compass and a hand-rubbed walnut steering wheel.

BLOCK FEATURES
All Corvairs had an automatic choke and aluminium cylinder heads.

TYRES
White sidewalls could be ordered for an extra $29.

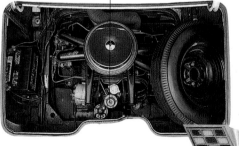

ENGINE
Corvair buyers had a choice of alloy, air-cooled, horizontal sixes. The base unit was a 164cid block with four Rochester carbs developing 140 bhp. The hot turbocharged motors were able to push out a more respectable 180 bhp.

POWER READING
The 140 badge represented the Corvair's power output.

END OF THE LINE

By the end of '68, sales of the handsome Monza Coupe were down to just 6,800 units and GM decided to pull the plug in May '69. Those who had bought a '69 Corvair were given a certificate worth $150 off any other '69–'70 Chevrolet.

SPECIFICATIONS

MODEL Chevrolet Corvair Monza (1966)

PRODUCTION 60,447 (1966, Monza only)

BODY STYLES Two- and four-door, four-seater coupé and convertible.

CONSTRUCTION Steel unitary body.

ENGINES 164cid flat sixes.

POWER OUTPUT 95–140 bhp.

TRANSMISSION Three-speed manual, optional four-speed manual, and two-speed Powerglide automatic.

SUSPENSION Front and rear coil springs.

BRAKES Front and rear drums.

MAXIMUM SPEED 169–193 km/h (105–120 mph)

0–60 MPH (0–96 KM/H) 11–15.2 sec

A.F.C. 7 km/l (20 mpg)

PRODUCTION

1965 model year production peaked at 205,000 units. Ford's Mustang did half a million in the same year.

FIRST MONZAS

The early Corvair Monzas, with de luxe trim and automatic transmission, were a big hit. In 1961, over 143,000 were sold, which amounted to over half the grand Corvair total.

WINDOWS

Side windows were made of specially curved glass.

AERIAL

Power-operated rear aerial was an option.

CHEVROLET *Camaro RS Convertible*

RUMOURS THAT GENERAL MOTORS had at last come up with something to steal sales from Ford's massively successful Mustang *(see pages 278–85)* swept through the American motor industry in the spring of 1966. Code-named Panther, the Camaro was announced to newspaper reporters on 29 June 1966, touching down in showrooms on 21 September. The Pony Car building-block philosophy was simple: sell a basic machine and allow the customer to add their own extras. The trouble was that the Camaro had an options list as arcane and complicated as a lawyer's library. From Strato-Ease headrests to Comfort-Tilt steering wheel, the Camaro buyer was faced with an *embarras de richesse*. But it worked. Buyers ordered the Rally Sport equipment package for their stock Camaros and suddenly they were kings of the street. Go-faster, twin-lined body striping, hidden headlamps, and matt black tail light bezels were all calculated to enhance the illusion of performance pedigree. Especially if he or she could not afford the real thing – the hot Camaro SS.

CLEAN STYLING
The market accepted the Camaro as a solid response to the Ford Mustang. Its styling was cleaner, more European, and less boxy, and it drove better than the Ford. Despite all this, Camaro sales were still considerably less than the Mustang.

PRODUCTION SOURCES
First-generation Camaros were mainly built at Norwood, Ohio, but some also came out of the Van Nuys plant in California.

NOSE JOB
The lengthened wheelbase created a big frontal overhang.

SPECIFICATIONS

MODEL Chevrolet Camaro RS Convertible (first generation, 1967–70)

PRODUCTION 10,675 (1967, RS), 195,765 (1967, coupé), and 25,141 (1967, convertible).

BODY STYLE Two-door, four-seater convertible.

CONSTRUCTION Steel monocoque.

ENGINE 327cid small block V8.

POWER OUTPUT 275 bhp at 4800 rpm.

TRANSMISSION Three- or four-speed manual, two- or three-speed auto.

SUSPENSION Independent front, rear leaf springs.

BRAKES Drums with optional power-assisted front discs.

MAXIMUM SPEED 177 km/h (110 mph)

0–60 MPH (0–96 KM/H) 8.3 sec

0–100 MPH (0–161 KM/H) 25.1 sec

A.F.C. 6.4 km/l (18 mpg)

RACING PEDIGREE
Chevy's Camaro was the chosen pace car for both the 1967 and '69 Indy 500s. Some of the production replicas were convertibles.

RAISING THE ROOF
The designers had created a sleek convertible – when the Camaro raised its roof, the purity of line was not disturbed.

REAR SPACE
GM liked to think that three passengers could be seated in the rear when in reality only two could be seated comfortably.

RS PINSTRIPING
Stick-on pinstriping helped flatter the Camaro's curves.

LIMITED STORAGE
For a car this big, the boot was incredibly small.

LIMITED NUMBERS
The Convertible RS was rare in 1967 with only 10,675 units produced.

SEATING
Strato-bucket front seats came as standard, but Strato-back bench seat could be specified as an extra.

RS REAR FEATURES
All-red tail light lenses with black bezels were an RS feature. Another part of the RS package was that reversing lights were moved to the rear valance panel. The RS emblem was inscribed on the fuel filler-cap.

COLOURED VINYL
Colour-keyed all-vinyl trim was a standard Camaro feature.

INTERIOR
Dash was the usual period fare, with acres of plastic and mock wood-grain veneer. This model is fitted with the optional four-speed manual gearbox.

RACING OPTION

Trans Am Racing spawned the Z28
Camaro, a thinly-veiled street racer,
designed to take on the Shelby Mustang.
Top speed was 200 km/h (124 mph)
and 0–60 came up in 6.7 seconds. Only
available as a coupé, it was designed for
those who put speed before comfort so
could not be ordered with automatic
transmission or air-conditioning.

ENGINE

The basic V8 power plant for Camaros was the
trusty small block cast-iron 327cid lump, which,
with a bit of tweaking, evolved into the 350cid unit
of the desirable SS models. Compression ratio
was 8.8:1 and it produced 210 bhp.

MIRROR CHANGE

*By 1968 the circular wing
mirrors had been replaced
by rectangular ones.*

POWER RATING

American horsepower was all about cubic
inches (cid), not cubic centimetres (cc) as
in Europe, and the RS proudly badged its
327 cubic inch capacity.

CHEVROLET *Corvette Stingray* (1969)

THE MOTORING PRESS REALLY LASHED into the '69 Shark, calling it a piece of junk, a low point in Corvette history, and the beginning of a new trend towards the image-and-gadget car. Instead of testing the 'Vette, *Car and Driver* magazine simply recited a litany of glitches and pronounced it "too dire to drive", sending ripples of rage through GM. To be frank, the '69 was not the best 'Vette ever. Styling was boisterous, boot space vestigial, the seats had you sliding all over the place, and the general build was shoddy. Two great engines saved the day, the 327cid and three incarnations of the big-block 427. With the hottest L88 version hitting 60 mph (96 km/h) in five-and-a-half seconds and peaking at 257 km/h (160 mph), these were cars that were race-ready from the showroom floor. Despite the vitriol, the public liked their image, gadgets, and grunt, buying 38,762 of them, a production record unbroken for the next six years – empirical proof that, occasionally, car journalists do talk hot air.

AGGRESSIVE POSTURE
The Stingray filled its wheelarches very convincingly with an aggressive, menacing presence. Any similarity to the European sports cars that inspired the original Corvettes had by now withered away, to be replaced by a new, threatening personality. In the annals of motoring history, there is no car with more evil looks than this 1968–72 generation Corvette.

STINGRAY BADGE
Chevy stopped calling their 'Vette the Sting Ray in 1968 but thought better of it in '69, reinstating the name as one word.

VENTILATION
Trim liners for side wing slots only appeared in the '69 model year.

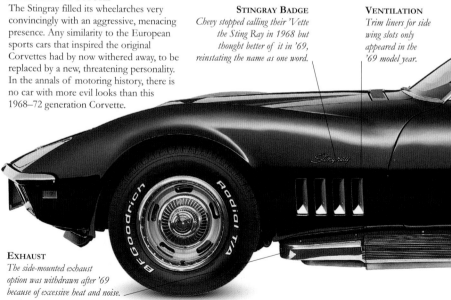

EXHAUST
The side-mounted exhaust option was withdrawn after '69 because of excessive heat and noise.

WINDOW
*Rear window
demister was
an option.*

RACK
*Rear rack helped as
there wasn't much
room in the boot.*

WHEELS
*Wheel-rim width
increased to 20 cm
(8 in) in 1969, wide
enough to roll an
English cricket pitch.*

RAD 'VETTE
*A four-wheel-drive,
mid-engined
prototype 'Vette
was developed but
cancelled in 1969.*

NEW DIRECTIONS

The '69 Stingray was styled by GM's Dave Hols and owed
little to the original Sting Ray. But this was the dawn of
the '70s and, while it might not have had the purest shape,
it reeked muscle from every vent.

AERIAL
*AM/FM radio option
was offered for the first
time in 1968.*

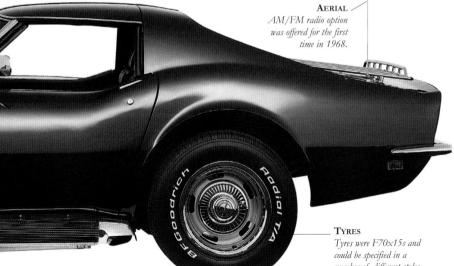

TYRES
*Tyres were F70x15s and
could be specified in a
number of different styles
including with white lettering.*

SHARK-BASED DESIGN

GM chief Bill Mitchell was an admirer of sharks – "they are exciting to look at"– and wanted to design a car with similar lines. In 1960 a prototype Mako Shark was made and the end result was the 1963 Sting Ray. A further prototype in 1966, the Mako Shark II, produced the 1968–72 generation of Stingray but the 'Vette collided with the energy crisis and would never be the same again.

ENGINE

If the stock 427 was not enough, there was always the 500 bhp ZL1, a 274 km/h (170 mph) racing option package. To discourage boy racers, no heater was installed in the ZL1; only two were ever sold to retail customers.

WINDSCREEN
Soft Ray tinted glass was an optional extra.

NOTABLE YEAR
1969 saw the 250,000th 'Vette come off the production line; it was a gold convertible.

BIG DADDY

With the 427 unit, the 'Vette was the biggest, heaviest, fastest, thirstiest, cheapest, and most powerful sports car on the market.

ENGINE OPTION
The first all-aluminium Corvette block was offered in 1969.

ROOF PANEL

Half of the '69 production were coupes with twin lift-off roof panels and a removable window — making this Stingray almost a convertible.

WIPER COVER

'68 and '69 'Vettes had a vacuum-operated lid which covered the windscreen wipers when not in use. It was, though, a styling gimmick which malfunctioned with depressing regularity.

INTERIOR

A major drawback of the '69 was its sharply raked seats, which prevented the traditional Corvette arm-out-of-the-window pose. While the telescopic tilt column and leather trim were extras, the glove compartment had been introduced as standard in 1968.

SPECIFICATIONS

MODEL Chevrolet Corvette Stingray (1969)

PRODUCTION 38,762 (1969)

BODY STYLES Two-seater sports and convertible.

CONSTRUCTION Glass-fibre, separate chassis.

ENGINES 327cid, 427cid V8s.

POWER OUTPUT 300–500 bhp.

TRANSMISSION Three-speed manual, optional four-speed manual, three-speed Turbo Hydra-Matic automatic.

SUSPENSION *Front:* upper and lower A-arms, coil springs; *Rear:* independent with transverse strut and leaf springs.

BRAKES Front and rear discs.

MAXIMUM SPEED 188–274 km/h (117–170 mph)

0–60 MPH (0–96 KM/H) 5.7–7.7 sec

A.F.C. 3.5 km/l (10 mpg)

HEADLIGHTS

The '69 retained hidden headlights, but now worked off a vacuum.

CHEVROLET *Monte Carlo*

NOW THE WORLD'S LARGEST PRODUCER of motor vehicles, Chevrolet kicked off
the Seventies with their Ford Thunderbird chaser, the 1970 Monte Carlo. Hailed
as "action and elegance in a sporty personal luxury package", it was only available as
a coupe and came with power front discs, Elm-Burl dash-panel inlays, and a choice of
engines that ranged from the standard 350cid V8 to the Herculean SS 454. At $3,123
in base form, it was cheap compared to the $5,000 needed to buy a Thunderbird. But
the T-Bird had become as urbane as Dean Martin and the Monte couldn't match the
Ford's élan. Even so, despite a six-week strike that lost Chevrolet 100,000 sales, over
145,000 Monte Carlos found buyers which, compared to a mere 40,000 T-Birds,
made Chevy's new personal luxury confection a monster hit.

SHARED CHASSIS
The Monte Carlo used the same
platform as the redesigned 1969
Pontiac Grand Prix. Stylistically, the
long bonnet and short boot promised
performance and power. The single
headlights were mounted in square-
shaped housings, and the grid-textured
grille was simple and unfussy.

INTERIOR
*The Monte Carlo's cabin was
Chevrolet's most luxurious for the
year, but was criticized for having
limited front and rear legroom.*

HIDDEN AERIAL
*The radio aerial was
hidden in the windscreen.*

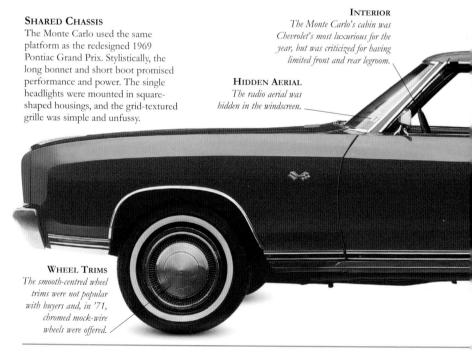

WHEEL TRIMS
*The smooth-centred wheel
trims were not popular
with buyers and, in '71,
chromed mock-wire
wheels were offered.*

SPEEDY UNIT
The massive 454 block made it a favourite with short-circuit stock car racers.

SPECIFICATIONS

MODEL Chevrolet Monte Carlo (1970)

PRODUCTION 145,975 (1970)

BODY STYLE Two-door, five-seater coupé.

CONSTRUCTION Steel body and chassis.

ENGINES 350cid, 400cid, 454cid V8s.

POWER OUTPUT 250–360 bhp.

TRANSMISSION Three-speed manual, optional two-speed Powerglide automatic, Turbo Hydra-Matic three-speed automatic.

SUSPENSION *Front:* coil springs; *Rear:* leaf springs.

BRAKES Front and rear drums.

MAXIMUM SPEED 185–211 km/h (115–132 mph)

0–60 MPH (0–96 KM/H) 8–14 sec

A.F.C. 5.3–7 km/l (15–20 mpg)

HEADLIGHTS
In '72, vertical parking lights were placed inboard of the headlights.

ENGINE
The potent SS 454 option was a modest $147 and could catapult the Monte Carlo to 60 mph (96 km/h) in less than eight seconds.

PILLAR
Prodigious rear pillar made city parking literally hit-or-miss.

VINYL ROOF
Black vinyl top was a $120 option. Buyers could also choose blue, dark gold, green, or white.

REAR STABILITY
Another option available, and fitted on this car, was rear anti-sway bars.

CHEVROLET *Nova SS*

THE NOVA NAME FIRST APPEARED in 1962 as the top-line model of Chevrolet's new Falcon-buster compact, the Chevy II. Evolving into a range in its own right, by '71 the Nova's Super Sport (SS) package was one of the smallest muscle cars ever fielded by Detroit. In an era when performance was on the wane, the diminutive banshee found plenty of friends among the budget drag-racing set. That strong 350cid V8 just happened to be a small-block Chevy, perfect for all those tweaky manifolds, carbs, headers, and distributors courtesy of a massive hop-up industry. Some pundits even went so far as hailing the Nova SS as the Seventies equivalent of the '57 Chevy. Frisky, tough, and impudent, Chevy's giant-killer could easily double the legal limit and the SS was a Nova to die for. Quick and rare, only 7,016 '71 Novas sported the magic SS badge. Performance iron died a death in '72, making these last-of-the-line '71s perfect candidates for the "Chevy Muscle Hall of Fame".

DIFFERENT STYLING
Handsome, neat, and chaste, the Nova was a new breed of passenger car for the Seventies. Advertised as the "Not Too Small Car", it looked a lot like a scaled-down version of the Chevelle and debuted in this form in 1968 to rave reviews.

SAFETY REFLECTORS
Side marker-lights were forced on the Nova after federal safety legislation was passed.

AIR-CON
*Air-conditioning was
an extra-cost option.*

LIGHTS
*Amber plastic
light lenses were
new for '71.*

INTERIOR
*Nova features included front
armrests, anti-theft steering-
wheel-column lock, and
ignition key alarm system.*

ENGINE
The two- or four-barrel 350cid V8
ran on regular fuel and pushed out
270 ponies. At one point, Chevrolet
planned to squeeze the massive
454cid V8 from the Chevelle into
the Nova SS, but regrettably
dropped the idea.

STYLING
*The Nova's shell would last for
11 years and was shared with
Buick, Oldsmobile, and Pontiac.*

BLOCK
*In '71, the option of a four-
cylinder block was withdrawn on
the Nova; less than one per cent
of '70 Nova buyers chose a four.*

ALLOYS
*The handsome
Sportmag five-
spoke alloys were
an $85 option.*

SPECIFICATIONS

MODEL Chevrolet Nova SS (1971)

PRODUCTION 7,016 (1971)

BODY STYLE Two-door, five-seater coupé.

CONSTRUCTION Steel unitary body.

ENGINE 350cid V8.

POWER OUTPUT 245 bhp.

TRANSMISSION Three-speed manual,
optional four-speed manual, and
three-speed automatic.

SUSPENSION *Front:* coil springs;
Rear: leaf springs.

BRAKES Front discs and rear drums.

MAXIMUM SPEED 193 km/h (120 mph)

0–60 MPH (0–96 KM/H) 6.2 sec

A.F.C. 7 km/l (20 mpg)

CHEVROLET *Camaro SS396*

AFTER A SUCCESSFUL DEBUT IN '67, the Camaro hit the deck in '72. Sluggish sales and a 174-day strike at the Lordstown, Ohio, plant meant Camaros were in short supply and only 68,656 were produced that year. Worse still, 1,100 half-finished cars sitting on the assembly lines couldn't meet the impending '73 bumper impact laws, so GM were forced to junk the lot. There were some dark mutterings in GM boardrooms. Should the Camaro be canned? 1972 also saw the Super Sport (SS) package bow out. *Road & Track* magazine mourned its passing, hailing the SS396 as "the best car built in America in 1971". But the early Seventies were a bad trip for the automobile, and the Camaro would rise again; five years later it had risen from the ashes and was selling over a quarter of a million units. This is one American icon that refuses to die.

DURABLE DESIGN

The Camaro design survived an incredible 11 years without any serious alteration. It lured eyes and dollars away from the traditional European performance machines and became one of the most recognized American GTs of the Seventies. As well as the SS package, Camaros could also be specified in Rally Sport (RS) and Z-28 performance guise.

SS NUMBERS

Only 6,562 Camaros had the SS equipment package in 1972 out of total Chevrolet sales for the year of 2,151,076.

STYLING

The Camaro was designed using computer technology; the smooth, horizontal surfaces blended together in an aerodynamically functional shape.

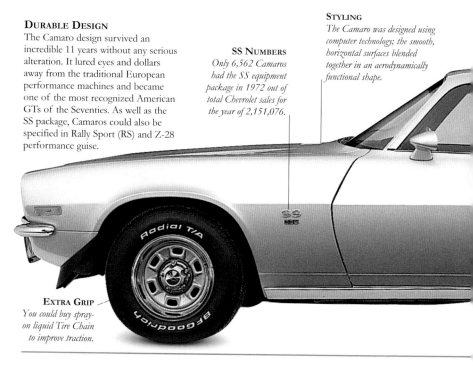

EXTRA GRIP

You could buy spray-on liquid Tire Chain to improve traction.

NASCAR RACER

Chevy spent big bucks to become performance heavyweights, and the Camaro, along with the Chevelle, was a successful racing model in the early '70s.

SPECIFICATIONS

MODEL Chevrolet Camaro SS396 (1972)

PRODUCTION 6,562 (SS, 1972)

BODY STYLE Two-door coupé.

CONSTRUCTION Steel body and chassis.

ENGINES 350cid, 396cid, 402cid V8s (SS).

POWER OUTPUT 240–330 bhp.

TRANSMISSION Three-speed manual, optional four-speed manual, and automatic.

SUSPENSION *Front:* coil springs; *Rear:* leaf springs.

BRAKES Front power discs and rear drums.

MAXIMUM SPEED 201 km/h (125 mph)

0–60 MPH (0–96 KM/H) 7.5 sec

A.F.C. 5.3 km/l (15 mpg)

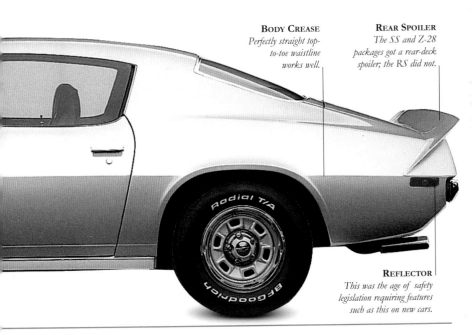

BODY CREASE
Perfectly straight top-to-toe waistline works well.

REAR SPOILER
The SS and Z-28 packages got a rear-deck spoiler; the RS did not.

REFLECTOR
This was the age of safety legislation requiring features such as this on new cars.

INTERIOR

Interiors were generally quite basic. Revisions for '72 were quite limited and mostly confined to the door panels – these now included map bins and coin holders under the door handles. The high-back seats are a clue that this is a post-'70 model.

COMFORT OPTIONS

Special instrumentation, centre console, and Comfort-Tilt wheel were convenience options.

UNIQUE SS

Unlike other performance packs, the SS option gave the car a whole new look. The bolt-on front end was different, and included sidelights up alongside the headlights and recessed grille. SS spec usually included mini quarter-bumpers rather than the full-width item seen here.

CONCEALED WIPERS

SS and RS packages included hidden windscreen wipers.

COOL INTERIOR
Air-conditioning for the Camaro cost an additional $397.

COMPUTER-DESIGNED
The Camaro was designed using computer technology, with smooth horizontal surfaces blended together in an aerodynamically functional shape. And individuality and power came cheap in '72 – the SS package cost just $306.

SS PANEL
The black rear panel was unique to the SS396.

SUPER BLOCK OPTION
The legendary 454cid V8, with a mind-blowing 425 bhp, was definitely not for the faint-hearted.

WHEELS
Camaros came with five wheel-trim options.

ENGINE
Camaros came with a range of engines to suit all pockets and for all types of drivers. The entry-level V8 was just $96 more than the plodding straight-six. The block featured here is the lively 396cid V8. Under 5,000 owners chose a six compared to nearly 64,000 who opted for one of the V8 options.

ENGINE IDEA
A 400cid engine was planned for mid-year introduction but it never made the Camaro.

CHRYSLER *Imperial*

IN 1950 CHRYSLER WERE CELEBRATING their silver jubilee, an anniversary year with a sting in its tail. The Office of Price Stabilization had frozen car prices, there was a four-month strike, and serious coal and steel shortages were affecting the industry. The '50 Imperial was a Chrysler New Yorker with a special roof and interior trim from the Derham Body Company. The jewels in Chrysler's crown, the Imperials were meant to lock horns with the best of Cadillac, Packard, and Lincoln. With Ausco-Lambert disc brakes, Prestomatic transmission, and a MoPar compass, they used the finest technology Chrysler could muster. The trouble was, only 10,650 Imperials drove out of the door in 1950, the hemi-head V8 wouldn't arrive until the next year, buyers were calling it a Chrysler rather than an Imperial, and that frumpy styling looked exactly like what it was – yesterday's lunch warmed up again.

BEASTS OF THE ROAD
Bulky, rounded Chryslers were some of the biggest cars on the road in 1950. The Imperials had Cadillac-style grilles, and the Crown Imperial was a long limousine built to rival the Cadillac 75.

WINDSCREEN
The front screen was still old-fashioned two-piece flat glass, which made the Imperial look rather antiquated.

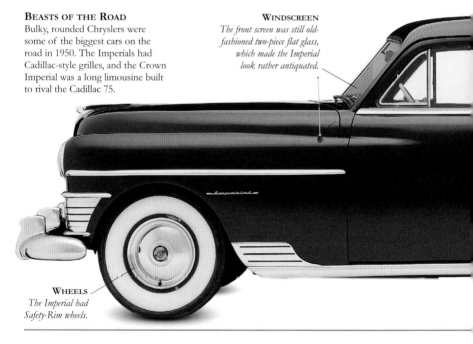

WHEELS
The Imperial had Safety-Rim wheels.

INTERIOR

Chrysler's interiors were as restrained and conservative
as the people who drove them. Turn-key ignition
replaced push-button in 1950, which was also
the first year of electric windows.

SPECIFICATIONS

MODEL Chrysler Imperial (1950)

PRODUCTION 10,650 (1950)

BODY STYLE Four-door sedan.

CONSTRUCTION Steel body and chassis.

ENGINE 323cid straight-eight.

POWER OUTPUT 135 bhp.

TRANSMISSION Prestomatic
semi-automatic.

SUSPENSION *Front:* coil springs;
Rear: live axle.

BRAKES Front and rear drums, optional
front discs.

MAXIMUM SPEED 161 km/h (100 mph)

0–60 MPH (0–96 KM/H) 13 sec

A.F.C. 5.7 km/l (16 mpg)

FUEL CAP
*The Imperial was
able to manage
5.7 km/l (16 mpg).*

LONGER WINGS
*Rear wings got longer
for 1950 and lights were
now nicely faired-in.*

BIGGER BLOCK
*180 bhp hemi-head
V8 wouldn't arrive
till next year.*

WASHERS
*Windscreen washers were
available as an option.*

ENGINE
The inline L-head eight developed
135 bhp and had a cast-iron block with
five main bearings. The carburettor was
a Carter single-barrel, and Prestomatic
automatic transmission with fluid
drive came as standard.

SUSPENSION
*Imperials incorporated
Safety-Level ride.*

SEMI-AUTOMATIC TRANSMISSION
The semi-automatic gearbox allowed
the driver to use a clutch to pull away,
with the automatic taking over as
the car accelerated. Imperials had
a waterproof ignition system.

LENGTH
*Wheelbase measured 334 cm
(131½ in), which was 36 cm (14 in)
shorter than the Crown Imperial.*

REAR SCREEN
*New "Clearbac" rear
window used three pieces
of glass that were
divided by chrome strips.*

LATE ARRIVAL
The celebrated designer Virgil Exner joined
Chrysler in 1949 but arrived too late to
improve the looks of the moribund Imperial.
Despite Chrysler's problems, 1950 was a
bumper year for American car production
with the industry wheeling out a
staggering 6,663,461 units.

TOP CAR
*Imperials were seen as the cream
of the Chrysler range. Advertising
for the Crown Imperial purred that
it was "the aristocrat of cars".*

IMPERIAL PRICING
The Imperial four-door sedan cost $3,055
before optional extras were added. The
most expensive model in Chrysler's 1950
range was the eight-passenger Crown
Imperial sedan, which cost $5,334. In
keeping with its establishment image, an
Imperial station wagon was never offered.
One claim to fame was that MGM Studios
used an Imperial-based mobile camera car
in many of their film productions.

WEIGHT
*The Imperial weighed just under
454 kg (1,000 lb) less than
the Crown Imperial.*

CHRYSLER *New Yorker*

WHY CAN'T THEY MAKE CARS that look this good anymore? The '57 New Yorker was the first and finest example of Chrysler's "Forward Look" policy. With the average American production worker earning $82.32 a week, the $4,259 four-door hardtop was both sensationally good-looking and sensationally expensive. The car's glorious lines seriously alarmed Chrysler's competitors, especially since the styling was awarded two gold medals, the suspension was by newfangled torsion bar, and muscle was courtesy of one of the most respected engines in the world – the hemi-head Fire Power. Despite this, "the most glamorous cars of a generation" cost Chrysler a whopping $300 million and sales were disappointing. One problem was a propensity for rust, along with shabby fit and finish; another was low productivity – only a measly 10,948 four-door hardtop models were produced. Even so, the New Yorker was certainly one of the most beautiful cars Chrysler ever made.

ONE MAN'S SHOW
Chrysler stunned the world with their dart-like shapes of 1957. The unified design was created by the mind of one man – Virgil Exner – rather than by a committee, and it shows. Those prodigious rear wings sweep up gracefully, harmonizing well with the gently tapering roof line.

MIRROR
Wing mirror was an optional extra.

AUTO FIRST
TorqueFlite automatic transmission was first seen this year.

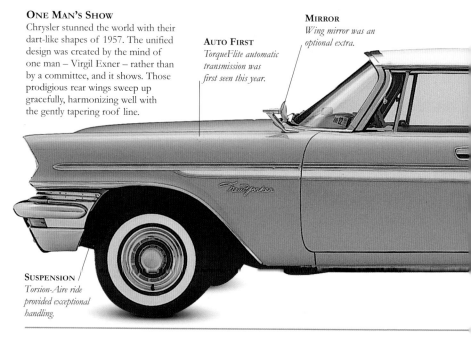

SUSPENSION
Torsion-Aire ride provided exceptional handling.

SPECIFICATIONS

MODEL Chrysler New Yorker (1957)

PRODUCTION 34,620 (all body styles, 1957)

BODY STYLE Four-door, six-seater hardtop.

CONSTRUCTION Monocoque.

ENGINE 392cid V8.

POWER OUTPUT 325 bhp.

TRANSMISSION Three-speed TorqueFlite automatic.

SUSPENSION *Front:* A-arms and longitudinal torsion bar; *Rear:* semi-elliptic leaf springs.

BRAKES Front and rear drums.

MAXIMUM SPEED 185 km/h (115 mph)

0–60 MPH (0–96 KM/H) 12.3 sec

A.F.C. 4.6 km/l (13 mpg)

SIMPLE AND EFFECTIVE
Rather than looking overstyled, the rear end and deck are actually quite restrained. The licence plate sits neatly in its niche, the tail pipes are completely concealed, the bumper is understated, and even the rear lights are not too heavy-handed.

WINNING SHAPE
The New Yorker's shape was so universally acclaimed that it was awarded two Grand Prix D'Honneur and two gold medals by the Industrial Designers Institute.

STYLISH ORNAMENTATION
The New Yorker had few styling excesses. Even the gratuitous slashes on the rear wing did not look over the top.

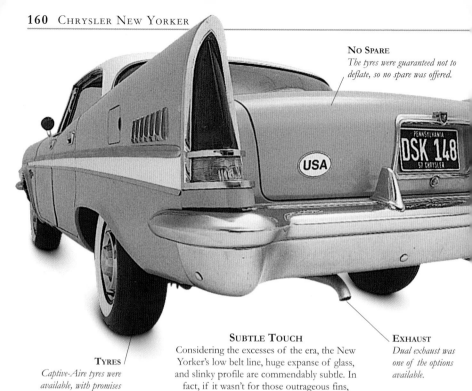

NO SPARE
The tyres were guaranteed not to deflate, so no spare was offered.

PENNSYLVANIA
DSK 148
57 CHRYSLER

USA

TYRES
Captive-Aire tyres were available, with promises that they wouldn't let themselves down.

SUBTLE TOUCH
Considering the excesses of the era, the New Yorker's low belt line, huge expanse of glass, and slinky profile are commendably subtle. In fact, if it wasn't for those outrageous fins, Chrysler's dreamboat might have ended up in the Museum of Modern Art.

EXHAUST
Dual exhaust was one of the options available.

INTERIOR
New Yorkers had the lot. Equipment included power windows, a six-way power seat, Hi-Way Hi-Fi phonograph, Electro-Touch radio, rear seat speaker, Instant Air heater, handbrake warning system, Air-Temp air-conditioning, and tinted glass – an altogether impressive array of features for a 1957 automobile. There are still many modern luxury cars that don't have the same comprehensive specification of the Fifties' New Yorker.

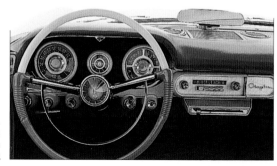

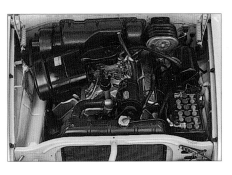

ENGINE

The top-of-the-range model had a top-of-the-range motor. The hemi-head was the largest production unit available in 1957. Bore and stroke were increased and displacement raised by nearly 10 per cent. It was efficient, ran on low-octane gas, and could be highly tuned.

OTHER MODELS

The three other model ranges for Chrysler in '57 were the Windsor, Saratoga, and 300C.

ESTATE VERSION

One of the other models in the 1957 New Yorker line-up was a Town and Country Wagon, which was driven by the same impressive Fire Power V8 found in the sedan and hardtops.

PENNSYLVANIA
DSK 148
57 CHRYSLER

Chrysler *300F* (1960)

"Red hot and rambunctious" is how Chrysler sold the 300F. It may be one of the strangest straplines of any American auto maker, but the 300F really was red hot and a serious flying machine that could better 225 km/h (140 mph). The rambunctious refers to the ram-air induction on the bad-boy 413cid wedge-head V8. Ram tuning had long been a way of raising torque and horsepower for drag racing, and it gave the 300F a wicked performance persona. One of Virgil Exner's happier designs, the 300F of '60 had unibody construction, a French Pont-A-Mousson four-speed gearbox, and front seats that swivelled towards you when you opened the doors. It also boasted an electro-luminescent instrument panel and Chrysler's best styling effort since 1957. But at $5,411, it was no surprise that only 964 coupés found buyers. Nevertheless, it bolstered Chrysler's image, and taught them plenty of tuning tricks for the muscle-car wars that were revving up just around the corner.

Power and Glory
The 300F was one of America's most powerful cars, and a tuned version recorded a one-way run of an amazing 304 km/h (189 mph) on the Bonneville salt flats. But despite the prodigious performance, it was deliberately understated compared with many contemporary Detroit offerings.

Non-inline Carbs
The alternative carburettor positioning gave a steady build-up of power along the torque curve.

Door Action
Opening the door initiated the self-activating swivelling seats.

Tyres
Nylon whitewalls came as standard.

NICKNAME
The phrase "beautiful brutes" was coined to describe the 300 Series.

FINE FINS
You could argue that the 300F's fins started at the front of the car and travelled along the side, building up to lethal, dagger-like points above the exquisitely sculptured tail lights.

PILLARLESS STYLE
With the window rolled down the 300F had a pillarless look.

LIMITED TIME
Within two years fins would disappear completely on the Chrysler letter series 300.

EXTRA GRIP
This particular model has Sure-Grip differential, a $52 option.

SPECIFICATIONS

MODEL Chrysler 300F (1960)

PRODUCTION 1,212 (1960, both body styles)

BODY STYLES Two-door coupé and convertible.

CONSTRUCTION Steel unitary body.

ENGINE 413cid V8.

POWER OUTPUT 375–400 bhp.

TRANSMISSION Three-speed push-button automatic, optional four-speed manual.

SUSPENSION *Front:* torsion bars; *Rear:* leaf springs.

BRAKES Front and rear drums.

MAXIMUM SPEED 225 km/h (140 mph)

0–60 MPH (0–96 KM/H) 7.1 sec

A.F.C. 4.2 km/l (12 mpg)

DASHBOARD

The "Astra-Dome" instrumentation was illuminated at night by electro-luminescent light, giving a soft, eerie glow that shone through the translucent markings on the gauges. It was technically very daring and boasted six different laminations of plastic, vitreous, and phosphor.

TINTED SCREEN

Solex tinted glass was a $43 optional extra.

TACHOMETER

Centre-mounted tachometer came as standard.

DANGER FINS

The 300F's razor-sharp rear fins were criticized by Ralph Nader in his book Unsafe at Any Speed *as "potentially lethal".*

THE ONLY BLEMISH

The much-criticized fake spare-tyre embellishment on the boot was variously described as a toilet seat or trash-can lid. This questionable rear deck treatment was officially known as "Flight-Sweep" and was also available on other Chryslers. Possibly the 300F's only stylistic peccadillo, it was dropped in '61.

QUIRKY SEAT SYSTEM
elf-activated swivelling seats
ere new for 1960 and pivoted
utwards automatically when
ther door was opened. It's
onic that the burly 300F's
pical owner was
eckoned to be a
abby 40-year-old.

AERIAL
*Power antenna was
a $43 option; this
car also has the
Golden Tone
radio ($124).*

MIRROR
*Wing mirror was
remote-controlled.*

SERIOUS STORAGE
The two-door shape meant that the rear deck was the size
of Indiana, and the cavernous boot was large enough to
hold four wheels and tyres.

Chrysler *300L (1965)*

Back in '55, Chrysler debuted their mighty 300 "Letter Car". The most powerful automobile of the year, the 300C kicked off a new genre of Gentleman's Hot-Rod that was to last for more than a decade. Chrysler cleverly flagged annual model changes with letters, running from the 300B in 1956 all the way through – the letter I excepted – to this 300L in 1965. And '65 was the swan-song year for the Letter Series speciality car. The 300L sat on high-performance rubber and suspension and was powered by a high-output 413cid 360 bhp mill breathing through a four-barrel Carter carb. By the mid-Sixties, though, the game had changed and Chrysler were pumping their money into muscle-car iron like the Charger and GTX, an area of the market where business was brisk. The 300L was the last survivor of an era when the Madison Avenue advertising men were still trying to persuade us that an automobile as long as a freight train could also be a sports car.

New Design Chief

Styling of the 300L was by Elwood Engle, who had replaced Virgil Exner as Chrysler's chief of design. Although the company's advertising claimed that this was "The Most Beautiful Chrysler Ever Built", the "Crisp, Clean, Custom" look of '63–'64 had ballooned.

Roomy Inside

Belt lines were lower and roof lines higher this year, which increased the glass area and made the interior feel even more cavernous.

Suspension

Torsion-bar front suspension gave poise and accuracy.

TOUGH JOB
Competition was particularly stiff in '65 and the 300L had to fight hard against the Oldsmobile Starfire, the agonizingly pretty Buick Riviera, and the market leader, Ford's flashy Thunderbird. Only 2,405 300L hardtops were produced, and a measly 440 two-door convertibles rolled out of the factory.

SPECIFICATIONS

MODEL Chrysler 300L (1965)

PRODUCTION 2,845 (1965)

BODY STYLES Two-door hardtop and convertible.

CONSTRUCTION Steel unitary body.

ENGINE 413cid V8.

POWER OUTPUT 360 bhp.

TRANSMISSION Three-speed automatic, optional four-speed manual.

SUSPENSION *Front:* torsion bar; *Rear:* leaf springs.

BRAKES Front and rear drums.

MAXIMUM SPEED 177 km/h (110 mph)

0–60 MPH (0–96 KM/H) 8.8 sec

A.F.C. 4.2–5 km/l (12–14 mpg)

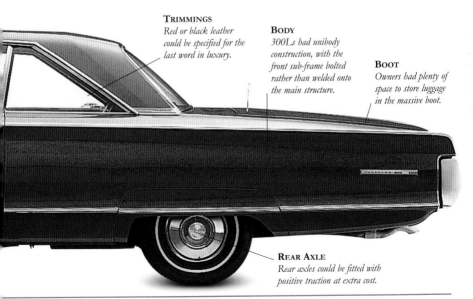

TRIMMINGS
Red or black leather could be specified for the last word in luxury.

BODY
300Ls had unibody construction, with the front sub-frame bolted rather than welded onto the main structure.

BOOT
Owners had plenty of space to store luggage in the massive boot.

REAR AXLE
Rear axles could be fitted with positive traction at extra cost.

COSTLY MOTOR
Coupés weighed in at a solid $4,090 with convertibles stickering at $4,545.

GRADUAL DEMISE

1961 saw the 300G, which was the last model to sport Exner's fins. The following year was arguably the start of the decline of the series and by the time the famous 300 nameplate had reached its final year, the spark had gone. The 300L was not as quick as its forebears and is the least special of Chrysler's limited editions.

HEADLIGHTS
These live behind a horizontally etched glass panel.

NEW BODY

In '65 the Chrysler line changed dramatically with a new corporate C-body shared with upmarket Dodges and the Plymouth Fury.

COMFORT EXTRAS

Options included tilting steering wheel, Golden Tone radio, cruise control, remote trunk release, high-speed warning system and air-conditioning.

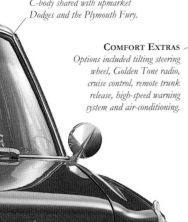

INTERIOR

Front bucket seats plus a centre console were standard on the L, as was the new-for-'65 column instead of push-button automatic gear shift. The rear seat was moulded to look like buckets but could actually accommodate three people.

ENGINE

The non-Hemi V8 was tough and reliable and gave the 300L very respectable performance figures. The L was quick, agile, and one of the smoothest-riding Letter Series cars made, with 45 bhp more than the standard 300's unit.

CITROËN *Traction Avant*

LOVED BY POLITICIANS, POETS, and painters alike, the Traction Avant marked a watershed for both Citroën and the world's motor industry. A design prodigy, it was the first mass produced car to incorporate a monocoque bodyshell with front-wheel drive and torsion bar springing, and it began Citroën's love affair with the unconventional. Conceived in just 18 months, the Traction Avant cost the French company dear. By 1934, they had emptied the company coffers, laid off 8,000 workers, and on the insistence of the French government, were taken over by Michelin, who gave the Traction Avant the backing it deserved. It ran for over 23 years, with over three quarters of a million saloons, fixed-head coupés, and cabriolets sold. Citroën's audacious saloon was the most significant and successful production car of its time, eclipsed only by the passage of 20 years and another *voiture revolutionnaire*, the Citroën DS.

WORLD-BEATER
With aerodynamic styling, unitary steel body, and sweeping wings without running boards, the Traction Avant was a technical and aesthetic *tour de force*.

FRONT WHEEL
Front-wheel drive made for tenacious roadholding.

BEAUTY NOT BRAWN
Though the Avant had a 1911cc engine, it only pushed out 46 bhp.

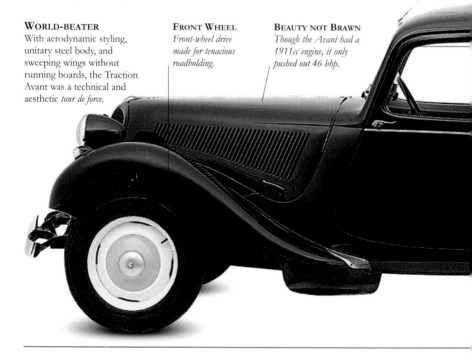

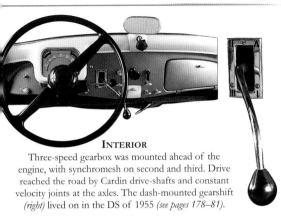

INTERIOR

Three-speed gearbox was mounted ahead of the engine, with synchromesh on second and third. Drive reached the road by Cardin drive-shafts and constant velocity joints at the axles. The dash-mounted gearshift *(right)* lived on in the DS of 1955 *(see pages 178–81).*

SPECIFICATIONS

MODEL Citroën Traction Avant (1934–55)

PRODUCTION 758,858 (including six-cylinder)

BODY STYLE Five-seater, four-door saloon.

CONSTRUCTION Steel front-wheel drive monocoque.

ENGINE 1911cc inline four-cylinder.

POWER OUTPUT 46 bhp at 3200 rpm.

TRANSMISSION Three-speed manual.

SUSPENSION Independent front and rear.

BRAKES Hydraulic drums front and rear.

MAXIMUM SPEED 113 km/h (70 mph)

0–60 MPH (0–96 KM/H) 25 sec

A.F.C. 8.1 km/l (23 mpg)

REVISED BOOT
In 1952, Citroën dispensed with the earlier "bob-tail" rear end and gave the Traction a "big boot".

WHEEL
Michelin produced these Pilote wheels and tyres for the Traction.

BONNET
Side-opening bonnet was a pre-war feature.

ENGINE
The Traction's Maurice Sainturat-designed engine was new. "Floating Power" came from a short-stroke four-cylinder unit, with a three-bearing crankshaft and push-rod overhead valves – equating to seven French horsepower.

EASY ACCESS
Engine, gearbox, radiator, and front suspension were mounted on a detachable cradle for easy maintenance.

STYLISH DESIGN
The Art Deco door handle is typical of Citroën's obsession with form and function. Beautiful yet practical, it epitomizes André Lefevre's astonishing design. The chevron-shaped gears were also pioneered for smoothness and silence.

TRICKY DRIVER
The Traction looks and feels huge and was a real handful in tight spaces.

REAR WINDOW
Small rear screen meant minimal rearward visibility.

SUSPENSION ATTRACTION

In 1954, as the car was approaching the end of its life, the six-cylinder Traction Avant was known as "Queen of the Road" because of its hydro-pneumatic suspension – a mixture of liquid and gas.

HOME COMFORTS
Citroën advertising tried to woo buyers with the line "on the road... the comfort of home".

FRONT SUSPENSION
All-independent suspension with torsion-bar springing, upper wishbones, radius arms, friction dampers, and worm-and-roller steering (later rack-and-pinion) gave crisp handling.

UNIVERSAL APPROVAL

The world lavished unstinting praise on the Traction Avant, extolling its roadholding, hydraulic brakes, ride comfort, and cornering abilities. Despite the praise, it was this great grand *routier* that devoured André Citroën's wealth and pushed him to his death bed.

REMOVABLE BONNET
Any serious engine repairs meant that the bonnet had to be removed completely.

CITROËN *2CV*

RARELY HAS A CAR BEEN SO ridiculed as the Citroën 2CV. At its launch at the 1948 Paris Salon, journalists lashed into this defenceless runabout with vicious zeal and everyone who was near Paris at the time claimed to be the originator of the quip, "Do you get a tin opener with it?" They all missed the point, for this minimal motor car was not meant to be measured against other motor cars; its true rival was the horse and cart, which Citroën boss Pierre Boulanger hoped to replace with his *toute petite voiture* – or very small car. As the Deux Chevaux it became much more than that and putt-putted into the history books, selling more than five million by the time of its eventual demise in 1990. As devotees of the 2CV say, "You either love them or you don't understand them".

CAREFUL PLANNING
In 1935, Pierre Boulanger conceived a car to woo farmers away from the horse and cart. It would weigh no more than 300 kg (661 lb), and carry four people at 60 km/h (37 mph), while consuming no more than 56 mpg. The car that appears "undesigned" was in fact carefully conceived.

SIMPLE CHASSIS
Although designers flirted with notions of a chassis-less car, cost dictated a more conventional sheet-steel platform chassis.

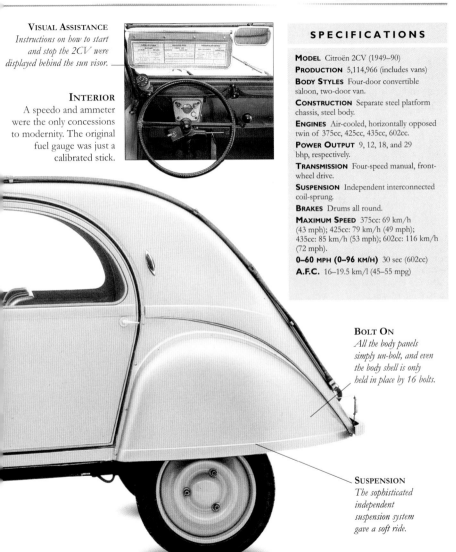

VISUAL ASSISTANCE
Instructions on how to start and stop the 2CV were displayed behind the sun visor.

INTERIOR
A speedo and ammeter were the only concessions to modernity. The original fuel gauge was just a calibrated stick.

SPECIFICATIONS

MODEL Citroën 2CV (1949–90)

PRODUCTION 5,114,966 (includes vans)

BODY STYLES Four-door convertible saloon, two-door van.

CONSTRUCTION Separate steel platform chassis, steel body.

ENGINES Air-cooled, horizontally opposed twin of 375cc, 425cc, 435cc, 602cc.

POWER OUTPUT 9, 12, 18, and 29 bhp, respectively.

TRANSMISSION Four-speed manual, front-wheel drive.

SUSPENSION Independent interconnected coil-sprung.

BRAKES Drums all round.

MAXIMUM SPEED 375cc: 69 km/h (43 mph); 425cc: 79 km/h (49 mph); 435cc: 85 km/h (53 mph); 602cc: 116 km/h (72 mph).

0–60 MPH (0–96 KM/H) 30 sec (602cc)

A.F.C. 16–19.5 km/l (45–55 mpg)

BOLT ON
All the body panels simply un-bolt, and even the body shell is only held in place by 16 bolts.

SUSPENSION
The sophisticated independent suspension system gave a soft ride.

STRAIGHTFORWARD DESIGN

The sober design purpose of the roll-top roof was to allow transportation of tall, bulky objects. It also happened that Citroën boss Pierre Boulanger was a six-footer who liked to wear a hat in a car. The minimal, but handy, lightweight, hammock-style seats lifted out to accommodate more goods or provide picnic seating.

BOOT
Roll-up canvas boot lid of the original saved both weight and cost; a metal boot lid took over in 1957 on French cars.

DOORS
You were lucky to get them; prototypes featured waxed-cloth door coverings.

537 BV-43
CITROËN

FUNCTIONAL DESIGN

The indicators are a good example of the functional design ethos. Why put a pair of indicators on the front and another pair on the back, when you could save the cost of two bulbs by giving your car cute "ears" that could be seen front and rear.

ENGINE

The original 375cc air-cooled twin, as seen here, eventually grew to all of 602cc, but all versions are genuinely happy to rev flat out all day. In fact, most spend all their time being driven at maximum speed and seem to thrive on full revs. Engines are hard-working and long-lasting.

AIR VENT
Fresh air was obtained by opening the vent on the scuttle; a mesh strained out the insects and leaves.

UNIQUE RIDE
Nothing drives like a Citroën 2CV – the handling looks lurid as it heels over wildly. The ride, though, is exceptional, and the tenacious grip of those skinny tyres is astonishing. All that and front-wheel drive too.

HEADLIGHT
Pre-war production prototypes had only one headlight.

BODY COLOURS
Grey until late 1959, then the choice doubled to include Glacier Blue, with green and yellow added in 1960.

Citroën *DS 21 Decapotable*

IN 1955, WHEN CITROËN FIRST drove prototypes of their mould-breaking DS through Paris, they were pursued by crowds shouting "La DS, la DS, voilà la DS!" Few other cars before or since were so technically and stylistically daring and, at its launch, the DS created as many column inches as the death of Stalin. Cushioned on a bed of hydraulic fluid, with a semi-automatic gearbox, self-levelling suspension, and detachable body panels, it rendered half the world's cars out of date at a stroke. Parisian coachbuilder Henri Chapron produced 1,365 convertible DSs using the chassis from the Safari Estate model. Initially Citroën refused to cooperate with Chapron but eventually sold the Decapotable models through their dealer network. At the time the swish four-seater convertible was considered by many to be one of the most charismatic open-top cars on the market, and today genuine Chapron cars command seriously hefty premiums over the price of "ordinary" tin-top DS saloons.

AERODYNAMIC PROFILE
The slippery, streamlined body cleaved the air with extreme aerodynamic efficiency. Body panels were detachable for easy repair and maintenance. Rear wings could be removed for wheel changing in minutes, using just the car's wheelbrace.

RENOWNED OWNERS
*Past owners of the DS include
General de Gaulle, Brigitte Bardot,
and the poet C. Day-Lewis.*

**THINNER
REAR**
*On all DSs
the rear track
was narrower
than the front.*

SPECIFICATIONS

MODEL Citroën DS 21 Decapotable (1960–71)

PRODUCTION 1,365

BODY STYLE Five-seater convertible.

CONSTRUCTION All-steel body with detachable panels, steel platform chassis with welded box section side members.

ENGINE Four-cylinder 2175cc.

POWER OUTPUT 109 bhp at 5550 rpm.

TRANSMISSION Four-speed clutchless semi-automatic.

SUSPENSION Independent all round with hydro-pneumatic struts.

BRAKES *Front:* disc; *Rear:* drums.

MAXIMUM SPEED 187 km/h (116 mph)

0–60 MPH (0–96 KM/H) 11.2 sec

0–100 MPH (0–161 KM/H) 40.4 sec

A.F.C. 8.5 km/l (24 mpg)

INTERIOR
The inside was as innovative as the outside, with clever use of curved glass and copious layers of foam rubber, even on the floors.

DASHBOARD
Bertone's asymmetrical dashboard makes the interior look as futuristic as the rest of the car. The single-spoke steering wheel was a Citroën hallmark. The dash-mounted gear lever operated the clutch-less semi-automatic box.

PROTECTION
Thin rubber over-rider-type bumpers offered some protection.

NOSE JOB
The DS was known as the "Shark" because of its prodigious nose.

QUALITY CHOICE

Smooth Bertone-designed lines have made the Citroën DS a cult design icon and the cerebral choice for doctors, architects, artists, and musicians. Customers could specify almost any stylistic or mechanical extra.

BADGING

Citroën's double chevrons are modelled on helical gears.

LIGHT ALTERATION

A major change came in 1967 when the headlamps and optional pod spot lamps were faired in behind glass covers.

ENGINE

The DS 21's rather sluggish 2175cc engine developed 109 bhp and was never highly praised, having its origins in the pre-war Traction Avant *(see pages 170–73)*. Stopping power was provided by innovative inboard disc brakes with split circuits.

SPARE WHEEL

Spare wheel under the bonnet allowed extra boot space.

SUSPENSION
*Fully independent
s suspension gave a
magic-carpet ride.*

STYLING
*Citroën's advertising
made much of the
car's futuristic looks.*

DS FAME
In 1962, the image of the DS received a boost
when terrorists attacked President General De
Gaulle. Despite being sprayed with bullets and
two flat tyres, the presidential DS was able to
swerve and speed away to safety.

NEAT TOUCHES
One of the Decapotable's
trademarks was angled chrome-
plated indicators perched on the
rear wings. Another was the novel
suspension, which could be raised
to clear rough terrain or navigate
flooded roads.

A TRUE CLASSIC
Low, rakish, and space-age in
appearance, the DS was so
perfectly styled that it hardly
altered shape in 20 years. The
French philosopher Roland
Barthes was captivated by the DS's
design and compared its technical
pre-eminence to the Gothic
flourish of medieval cathedrals.

CITROËN *SM*

THE CITROËN SM MAKES about as much sense as Concorde, but since when have great cars had anything to do with common sense? It is certainly a flight of fancy, an extravagant, technical *tour de force* that, as a 4.9-m (16-ft) long streamliner, offered little more than 2+2 seating. The SM bristled with innovations – many of them established Citroën hallmarks – like swivelling headlights and self-levelling hydro-pneumatic suspension. It was a complex car – too complex in fact, with self-centring power steering and brakes that were both powered by (and virtually inoperable without) a high-compression engine-driven pump. And of course there was that capricious Maserati V6 motor. Yet once again Citroën had created an enduringly futuristic car where other "tomorrow cars of today" were soon exposed as voguish fads.

SLEEK AND SPEEDY
The SM's striking low-drag body was designed by ex-General Motors stylist Henri de Segur Lauve. The sleek nose and deep undertray, together with the noticeably tapered rear end, endow it with a slippery profile that gives a high level of aerodynamic efficiency. It is impressively stable at high speed.

COMPOUND CURVES
The tinted rear window, with compound curves and heating elements, must have cost a fortune to produce.

US INFLUENCE
Only the SM's over-elaborate chromed rear "fins" betray the General Motors styling influence.

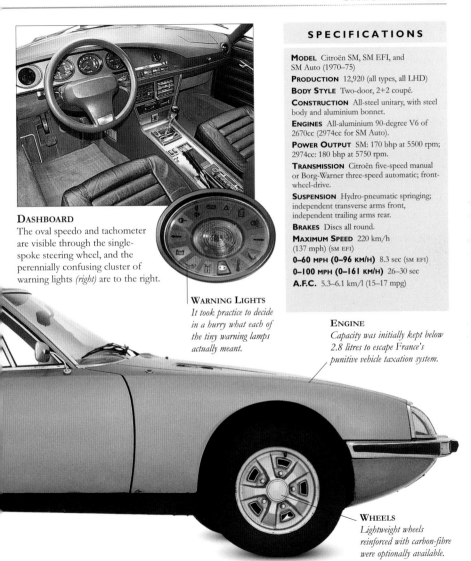

DASHBOARD
The oval speedo and tachometer are visible through the single-spoke steering wheel, and the perennially confusing cluster of warning lights *(right)* are to the right.

WARNING LIGHTS
It took practice to decide in a hurry what each of the tiny warning lamps actually meant.

SPECIFICATIONS

MODEL Citroën SM, SM EFI, and SM Auto (1970–75)

PRODUCTION 12,920 (all types, all LHD)

BODY STYLE Two-door, 2+2 coupé.

CONSTRUCTION All-steel unitary, with steel body and aluminium bonnet.

ENGINES All-aluminium 90-degree V6 of 2670cc (2974cc for SM Auto).

POWER OUTPUT SM: 170 bhp at 5500 rpm; 2974cc: 180 bhp at 5750 rpm.

TRANSMISSION Citroën five-speed manual or Borg-Warner three-speed automatic; front-wheel-drive.

SUSPENSION Hydro-pneumatic springing; independent transverse arms front, independent trailing arms rear.

BRAKES Discs all round.

MAXIMUM SPEED 220 km/h (137 mph) (SM EFI)

0–60 MPH (0–96 KM/H) 8.3 sec (SM EFI)

0–100 MPH (0–161 KM/H) 26–30 sec

A.F.C. 5.3–6.1 km/l (15–17 mpg)

ENGINE
Capacity was initially kept below 2.8 litres to escape France's punitive vehicle taxation system.

WHEELS
Lightweight wheels reinforced with carbon-fibre were optionally available.

SURPRISING HANDLING

Despite its size and weight, the SM can actually be thrown around like a sports car. It rolls like a trawler in a heavy sea and, like all front-wheel drivers, it understeers strongly but resolutely refuses to let go.

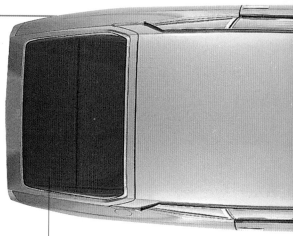

NOVEL LIGHTS

The SM had an array of six headlights, with the inner light on each side swivelling as the steering was turned.

WIND CHEATER

The tapering body is apparent in this overhead view.

PURELY FUNCTIONAL

The bulge in the tailgate above the rear number plate was for purely functional, aerodynamic reasons. It also suited the deeper licence plates used on models in the US.

SUPPORTING ROLE

Like that of most front-wheel drive cars, the SM's rear suspension did little more than hold the body off the ground.

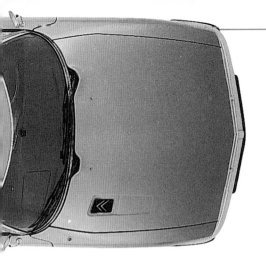

ENGINE
SM stands for Serié Maserati, and the exquisite Maserati all-aluminium V6 engine weighed just 140 kg (309 lb), was only 31 cm (12 in) long, but produced at least 170 bhp.

REAR CRAMP
Citroën's publicity material tried to hide the fact, but rear-seat legroom and headroom were barely sufficient for two large children.

FRONTWARD VISIBILITY
Slim windscreen pillars should have meant excellent visibility but, in practice, the left-hand drive SM was sometimes difficult to place on the road.

BRAKES
Inboard front disc-brakes incorporated the handbrake mechanism.

CONTINENTAL *Mark II*

THAT THE FIFTIES MOTOR INDUSTRY couldn't make a beautiful car is robustly disproved by the '56 Continental. As pretty as anything from Italy, the Mark II was intended to be a work of art and a symbol of affluence. William Ford was fanatical about his personal project, fighting for a chrome rather than plastic bonnet ornament costing $150, or the price of an entire Ford grille. But it was that tenacious attention to detail that killed the car. Even with the Mark II's huge $10,000 price tag, the Continental Division still haemorrhaged money. Poor sales, internal company struggles, and the fact that it was only a two-door meant that by '58 the Continental was no more. Ironically, one of the most beautiful cars Ford ever made was sacrificed to save one of the ugliest in the upcoming E-Car project – the Edsel.

PERSONAL LUXURY

The most expensive automobile in America, the $9,695 Continental really was the car for the stars. Elvis tried one as a change from his usual Cadillacs, and Jayne Mansfield owned a pearl-coloured '57 with mink trim. The Continental was three years in the planning and was sold and marketed through a special Continental Division.

BODY HEIGHT
"Cow belly" frame was specifically designed to allow high seating with a low roof line.

LEATHER INTERIOR
The high-quality all-leather trim was specially imported from Bridge of Weir in Scotland.

RAG-TOPS
Two special convertibles were built before the Continental was axed.

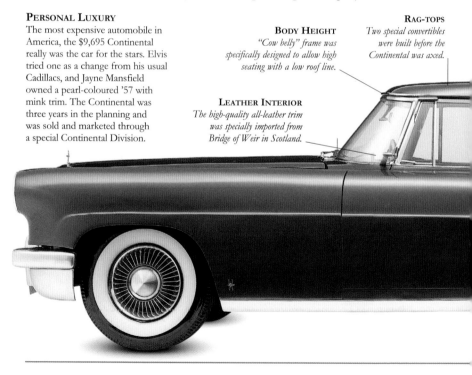

INTERIOR

The classically simple cockpit could have come straight out of a British car. The interior boasted richly grained leathers and lavish fabrics. Self-tuning radio, four-way power seat, dual heater, and map lights were among an impressive array of standard features.

SPECIFICATIONS

MODEL Continental Mark II (1956)

PRODUCTION 2,550 (1956)

BODY STYLE Two-door, four-seater sedan.

CONSTRUCTION Steel body and chassis.

ENGINE 368cid V8.

POWER OUTPUT 300 bhp.

TRANSMISSION Turbo-Drive three-speed automatic.

SUSPENSION *Front:* independent coil springs; *Rear:* leaf springs.

BRAKES Front and rear drums.

MAXIMUM SPEED 185 km/h (115 mph)

0–60 MPH (0–96 KM/H) 12.1 sec

A.F.C. 5.7 km/l (16 mpg)

CABIN TEMPERATURE
Air-conditioning was the Continental's only extra-cost option.

PRICE OF CLAY
A pricey little number, even the Continental's prototype clay mock-up cost $1 million.

SEATS
Seats were one of the many power-assisted elements of the car.

GAS GUZZLER
Like all US cruisers of the era, the Continental was a thirsty beast, with a figure of 5.7 km/l (16 mpg).

HANDSOME REAR
Handsome three-quarter profile echoes some Ferrari 250 models. Note how the petrol tank-cap lives behind the tail light. Unlike later models, the stamped-in spare tyre cover did actually house the spare.

BIG BLOCK
Excepting Packard's 374cid unit, this was the largest engine available in a 1956 production car.

ENGINE
Engines were Lincoln 368cid V8s, specially picked from the assembly line, stripped down, and hand-balanced for extra smoothness and refinement.

TINTED GLASS
This was one of the no-cost extras offered. Others included two-tone paint and an engraved nameplate.

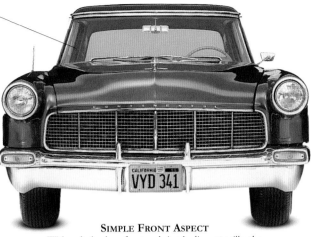

SCRIPT
Continental tag revived the famous 1930s Lincolns of Edsel Ford.

SIMPLE FRONT ASPECT
With a sleek, clean front and simple die-cast grille, the only concession to contemporary Detroit ornamentation was how the direction indicators were faired into the front bumper.

ROLLS-KILLER
At the rear of the car, trim fins, elegant bumpers, and neat inset tail lights meant that the Continental was admired on both sides of the Atlantic. But though its target market was Rolls-Royce territory, it turned out that the market wasn't large enough to sustain volume production.

CLASSY BODY
High-quality bodies were specially finished by the Mitchell-Bentley Corporation of Ionia, Michigan.

FRENCH DEBUT
The Continental debuted on 6 October 1955 at the Paris Auto Show to rave reviews.

DAIMLER *SP250 Dart*

AN ECCENTRIC HYBRID, the SP250 was the car that sunk Daimler. By the late Fifties, the traditionalist Coventry-based company was in dire financial straits. Hoping to woo the car-crazy Americans, Daimler launched the Dart, with its odd pastiche of British and American styling themes, at the 1959 New York Show. Daimler had been making buses out of glass-fibre and the Dart emerged with a quirky, rust-free glass-reinforced-plastic body. The girder chassis was an unashamed copy of the Triumph TR2 *(see pages 444–47)* and, to keep the basic price down, necessities like heater, windscreen washers, and bumpers were made optional extras. Hardly a great car, the SP250 was a commercial failure and projected sales of 7,500 units in the first three years dissolved into just 2,644, with only 1,200 going Stateside. Jaguar took over Daimler in 1960 and, by 1964, Sir William Lyons had axed the sportiest car Daimler had ever made.

THE DART CONCEPT
The Dart was a Fifties concept born too late to compete with the New Wave of monocoque sports cars headed by the stunning E-Type. It stands as a memorial to both the haphazard Sixties British motor industry and its self-destructive love affair with all things American.

BONNET
Glass-fibre bonnet had a nasty habit of springing open at speed.

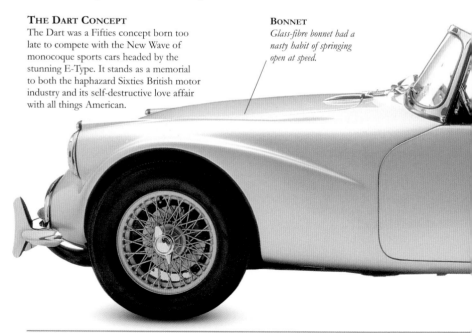

INTERIOR

The cockpit was pure British trad, with centre gauges mounted on an aluminium plate, leather seats and dash, an occasional rear seat, fly-off handbrake, wind-up windows, and thick-pile carpets. Borg-Warner automatic transmission was an option but tended to slow the car down considerably.

REAR SEAT
Vestigial rear seat could just about accommodate one child.

SPECIFICATIONS

MODEL Daimler SP250 Dart (1959–64)

PRODUCTION 2,644 (1,415 LHD, 1,229 RHD)

BODY STYLE Two-door, two-seater sports convertible.

CONSTRUCTION Glass-fibre body, steel girder chassis.

ENGINE Iron-block 2548cc V8.

POWER OUTPUT 140 bhp at 5800 rpm.

TRANSMISSION Four-speed manual or three-speed Borg-Warner Model 8.

SUSPENSION Independent front with wishbones and coil springs. Rear live axle with leaf springs.

BRAKES Four-wheel Girling discs.

MAXIMUM SPEED 201 km/h (125 mph)

0–60 MPH (0–96 KM/H) 8.5 sec

0–100 MPH (0–161 KM/H) 19.1 sec

A.F.C. 8.8 km/l (25 mpg)

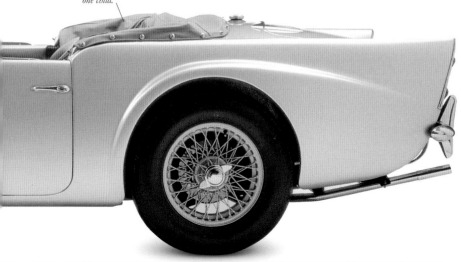

ENGINE

The turbine-smooth, Edward Turner-designed V8 was the Dart's *tour de force*. If you were brave enough, it could reach 201 km/h (125 mph). With alloy heads and hemispherical combustion chambers, it was a gem of a unit that survived until 1969 in the Daimler 250 saloon.

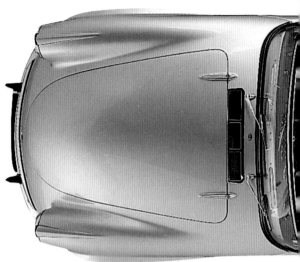

IMPOSING SIGHT

The guppy-style front could never be called handsome but, when Sixties drivers caught it in their rear-view mirrors, they knew to move over. The drastic plastic Dart was seriously quick. Contemporary tests praised the Dart's performance and sweet-running V8.

WINGS

Fluted wings look good and gave the body extra rigidity.

4068 WK

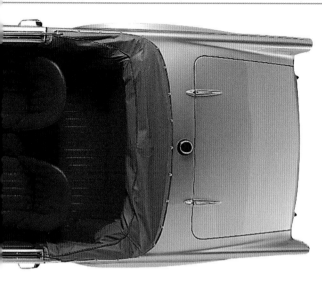

DART DEVELOPMENT

Dart development had three phases: 1959–61 A-spec cars came with no creature comforts; April 1961 and later B-specs had standard bumpers, windscreen washers, and chassis modifications; while the last and most refined C-specs, produced from April 1963 to September 1964, boasted a heater and cigar lighter as standard.

NEAT HOOD
Hood furled away neatly behind rear seat, covered with a fabric hood bag.

PEED STRAIN
t speed, the Dart was hard work;
e chassis flexed, doors opened on
ends, and the steering was heavy.
oad-testers admired
s speed but
ought the chassis,
andling, and body
nish were poor.

CUTE STYLING
*Chrome-on-brass
rear light finishers
ere monogrammed
th a dainty "D".*

4068 WK

DATSUN *Fairlady 1600*

THE SIMILARITY BETWEEN THE Datsun Fairlady and the MGB *(see pages 372–73)* is quite astonishing. The Datsun actually appeared first, at the 1961 Tokyo Motor Show, followed a year later by the MGB. Hardly a great car in its early 1500cc guise, the Fairlady improved dramatically over the years, a foretaste of the Japanese car industry's culture of constant improvement. The later two-litre, twin-carb, five-speed variants of 1967 could reach 200 km/h (125 mph) and even raised eyebrows at American sports car club races. Aimed at the American market, where it was known as the Datsun 1500, the Fairlady sold only 40,000 in nine years. But it showed Datsun how to make the legendary 240Z *(see pages 196–99)*, which became one of the world's best-selling sports cars.

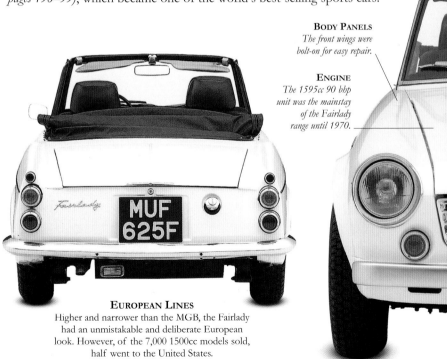

BODY PANELS
The front wings were bolt-on for easy repair.

ENGINE
The 1595cc 90 bhp unit was the mainstay of the Fairlady range until 1970.

EUROPEAN LINES
Higher and narrower than the MGB, the Fairlady had an unmistakable and deliberate European look. However, of the 7,000 1500cc models sold, half went to the United States.

STYLING

Interestingly, no attempt was made to make the interior harmonize with the Fairlady's traditional exterior lines. The cockpit was typical of the period, with acres of black plastic.

SPECIFICATIONS

MODEL Datsun Fairlady 1600 (1965–70)

PRODUCTION Approx 40,000

BODY STYLE Two-seater sports convertible.

CONSTRUCTION Steel body mounted on box-section chassis.

ENGINE 1595cc four-cylinder.

POWER OUTPUT 90 bhp at 6000 rpm.

TRANSMISSION Four-speed all-synchro.

SUSPENSION *Front:* independent; *Rear:* leaf springs.

BRAKES Front wheel discs, rear drums.

MAXIMUM SPEED 169 km/h (105 mph)

0–60 MPH (0–96 KM/H) 13.3 sec

0–100 MPH (0–161 KM/H) 25 sec

A.F.C. 8.8 km/l (25 mpg)

PERIOD CHARM

Low and rakish with classically perfect proportions, the Fairlady has a certain period charm and is one of the best-looking Datsuns produced before 1965. Side-on views show the car at its best, while the messy rear and cluttered nose do not work quite so well.

DATSUN *240Z*

THROUGHOUT THE 1960S, Japanese car makers were teetering on the brink of a sports car breakthrough. Toyota's 2000 GT *(see pages 442–43)* was a beauty but, with only 337 made, it was an exclusive curio. Honda was having a go too, with the dainty S600 and S800. As for Datsun, the MGB-lookalike Fairladies were relatively popular in Japan and the United States, but virtually unknown elsewhere. The revolution came with the Datsun 240Z, which at a stroke established Japan on the world sports car stage at a time when there was a gaping hole in that sector, particularly in the US. It was even launched in the States in October 1969, a month before its official Japanese release, and on a rising tide of Japanese exports to the US it scored a massive hit. It had the looks, performance, handling, and equipment levels. A great value sporting package that outsold all rivals.

TOP STYLIST
The lines of the 240Z were based on earlier styling exercises by Albrecht Goertz, master stylist of the BMW 507 *(see pages 64–67).*

SPOILER
Boot-lid aerofoil was not a standard 240Z fitment in all markets.

SCREEN
Steeply raked windscreen aided aerodynamic efficiency.

BALANCE
This view shows that the engine was placed forward of the centre-line, with the occupants well behind it; yet the Z was noted for its fine balance. The large rear window offered the driver excellent rearward vision.

BONNET
Bonnet was uncluttered by unnecessary louvres; it later became fussier.

ENGINE
The six-cylinder twin-carb 2.4-litre engine was developed from the four-cylinder unit of the Bluebird saloon range.

WHEELS
Tacky plastic wheel-trim is an original fitment.

FIRST-OF-BREED

As with so many long-lived sports cars, the first-of-breed 240Z is seen as the best sporting package – lighter and nimbler than its successors. If you wanted to cut a real dash in a 240Z, the ultimate Samurai performance option had what it takes. Modifications gave six-second 0–60 (96 km/h) figures.

MIXED STYLING CUES

As with the recessed lights at the front, there is an echo of the E-Type Jaguar fixed-head coupé *(see pages 306–09)* at the rear, with a little Porsche 911 *(see pages 420–21)*, Mustang fastback *(see pages 282–85)*, and Aston Martin DBS of 1969.

CAT LIGHTS
Recessed front light treatment is very reminiscent of an E-Type Jaguar.

INTERIOR

Cockpit layout was tailored to American tastes, with hooded instruments and beefy controls. The vinyl-covered bucket seats offered generous rear luggage space.

Z IDENTITY

The model was launched in Japan as the Fairlady Z, replacing the earlier Fairlady range; export versions were universally known as 240Z and badged accordingly. Non-UK and US models were badged as Nissans rather than Datsuns.

SPECIFICATIONS

MODEL Datsun 240Z (1969–73)

PRODUCTION 156,076

BODY STYLE Three-door, two-seater sports hatchback.

CONSTRUCTION Steel monocoque.

ENGINE Inline single overhead-camshaft six, 2393cc.

POWER OUTPUT 151 bhp at 5600 rpm.

TRANSMISSION All-synchromesh four- or five-speed manual gearbox, or auto.

SUSPENSION *Front*: Independent by MacPherson struts, low links, coil springs, telescopic dampers; *Rear*: Independent by MacPherson struts, lower wishbones, coil springs, telescopic dampers.

BRAKES Front discs, rear drums.

MAXIMUM SPEED 210 km/h (125 mph)

0–60 MPH (0–96 KM/H) 8.0 sec

A.F.C. 7–9 km/l (20–25 mpg)

BODY PANELS
Thin, rot-prone body panels were one of the few things that let the 240Z down.

BADGING
The name Datsun – literally son of Dat – first appeared on a small Dat in 1932.

SUSPENSION
Sophisticated suspension spec was independent with MacPherson struts on all four wheels.

RHX 156L

DeLorean *DMC 12*

"The long-awaited transport revolution has begun" bellowed the glossy brochures for John Zachary DeLorean's mould-breaking DMC 12. With a unique brushed stainless-steel body, gullwing doors, and an all-electric interior, the DMC was intended as a glimpse of the future. Today its claim to fame is as one of the car industry's greatest failures, on a par with Ford's disastrous Edsel *(see pages 216–23)*. Despite £65m worth of government aid to establish a purpose-built factory in West Belfast, DeLorean shut its doors in 1982 with debts of £25m. As for the hapless souls who bought the cars, they were faced with a litany of quality control problems, from doors that would not open, to windows that fell out. Even exposure in the film *Back to the Future* did not help the DeLorean's fortunes. Success depended on American sales and the company's forecasts were wildly optimistic. After the initial novelty died down, word spread that DeLoreans were dogs and sales completely evaporated.

BACHELOR WHEELS
The DeLorean was targeted at "the bachelor who's made it" and part of the design brief was that there had to be room behind the front seats for a full set of golf clubs. It was designed by Giugiaro and overseen by Colin Chapman of Lotus fame.

WHEELS
Custom-made spoked alloys were smaller at the front than the back.

LIGHT FRONT
With rear-engined layout, the weight distribution was split 35 per cent front to 65 per cent rear.

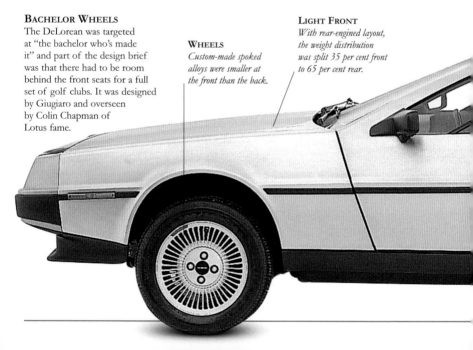

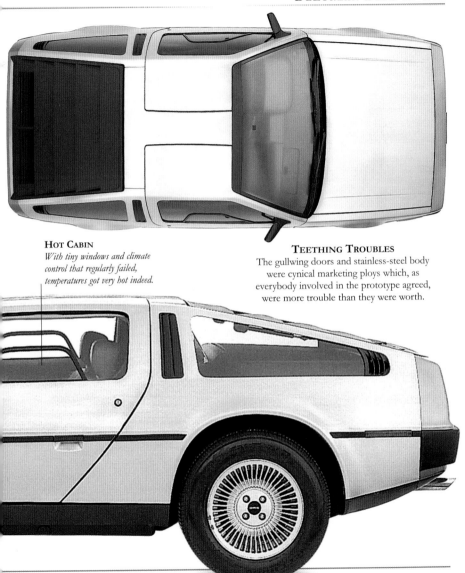

HOT CABIN
With tiny windows and climate control that regularly failed, temperatures got very hot indeed.

TEETHING TROUBLES
The gullwing doors and stainless-steel body were cynical marketing ploys which, as everybody involved in the prototype agreed, were more trouble than they were worth.

STARRING ROLE
The 1985 film *Back to the Future* used a DeLorean as a time machine to travel back to 1955; in reality the car was very orthodox. Underpinnings were technically uninspiring and relied heavily on components from other cars. Under the bonnet, the 145 bhp output was modest.

GULLWINGS
The DeLorean's most celebrated trick was gullwing doors that leak and did not open or close properly.

STRUT
Held by a puny single gas strut, it was an act of the purest optimism to expect the doors to work properly.

DATED DELOREAN
By the time of its launch in 1979, the DeLorean was old before its time. '70s styling motifs abound, like the slatted rear window and cubed rear lights.

ENGINE
The overhead-cam, Volvo-sourced 2.8 V6 engine used Bosch K-Jetronic fuel injection. Five-speed manual was standard with three-speed automatic optional.

ELECTRICS
Complex electrics were the result of last-minute cost-cutting measures.

HEAVY DOORS
Overloaded doors were crammed with locks, glass, electric motors, mirrors, stereo speakers, and ventilation pipery.

STAINLESS-STEEL BODY
Brushed stainless-steel was disliked by Colin Chapman but insisted upon by DeLorean himself. Soon owners found that it was impossible to clean.

INTERIOR
The leather-clad interior looked imposing, with electric windows, tilting telescopic steering column, double weather seals, air-conditioning, and a seven-position climate control function.

SPECIFICATIONS

MODEL DeLorean DMC 12 (1979–82)

PRODUCTION 6,500

BODY STYLE Two-seater rear-engined sports coupé.

CONSTRUCTION Y-shaped chassis with stainless-steel body.

ENGINE 2850cc ohc V6.

POWER OUTPUT 145 bhp at 5500 rpm.

TRANSMISSION Five-speed manual (optional three-speed auto).

SUSPENSION Independent with unequal length parallel arms and rear trailing arms.

BRAKES Four-wheel discs.

MAXIMUM SPEED 201 km/h (125 mph)

0–60 MPH (0–96 KM/H) 9.6 sec

0–100 MPH (0–161 KM/H) 23.2 sec

A.F.C. 7.8 km/l (22 mpg)

DeSoto *Custom*

The DeSoto of 1950 had a glittery glamour that cheered up post-war America. Hailed as "cars built for owner satisfaction", they were practical, boxy, and tough. DeSoto was a long-time taxi builder that, in the steel-starved years of 1946–48, managed to turn out 11,600 cabs, most of which plied the streets of New York. Despite more chrome upfront than any other Chrysler product, DeSotos still laboured on with an L-head six-pot 250cid mill. The legendary Firedome V8 wouldn't arrive until 1952. But body shapes for 1950 were the prettiest ever, and the American public reacted with delight, buying up 133,854 units in the calendar year, ranking DeSoto 14th in the industry. Top-line Custom Convertibles had a very reasonable sticker price of $2,578 and came with Tip-Toe hydraulic shift with Gyrol fluid drive as standard. The austere post-war years were a sales Disneyland for the makers of these sparkling cars, but DeSoto's roll couldn't last. By 1961 they'd disappeared forever.

MODEL RANGE
The top-of-the-line Custom range fielded a Club Coupé, two huge wagons, a six-passenger sedan, a two-door Sportsman, and a convertible. DeSoto's volume sellers were its sedans and coupés, which listed at under $2,000 in De Luxe form.

INNOVATIVE GEARING
Fluid drive gearbox was an innovative semi-automatic pre-selector with conventional manual operation or semi-auto kick-down.

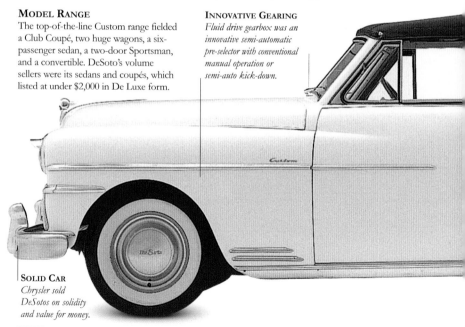

SOLID CAR
Chrysler sold DeSotos on solidity and value for money.

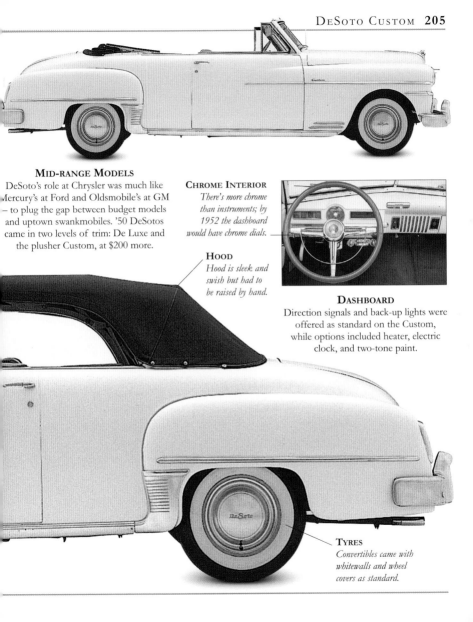

MID-RANGE MODELS

DeSoto's role at Chrysler was much like Mercury's at Ford and Oldsmobile's at GM – to plug the gap between budget models and uptown swankmobiles. '50 DeSotos came in two levels of trim: De Luxe and the plusher Custom, at $200 more.

CHROME INTERIOR
There's more chrome than instruments; by 1952 the dashboard would have chrome dials.

HOOD
Hood is sleek and swish but had to be raised by hand.

DASHBOARD
Direction signals and back-up lights were offered as standard on the Custom, while options included heater, electric clock, and two-tone paint.

TYRES
Convertibles came with whitewalls and wheel covers as standard.

REAR WING
The DeSoto body shape still carried hints of the separate wings of pre-war cars.

SPLIT SCREEN
Flat glass split screen was parted with a chromed centre rod on which the rear-view mirror was positioned.

CHUNKY YET REFINED
The DeSoto's rump was large, round, and unadorned and boot space was cavernous. The Custom Convertible was clean and elegant enough to be seen cruising along the smartest boulevards.

MASCOT
Optional bonnet mascot was one Hernando DeSoto, a 17th-century Spanish conquistador. The mascot glowed in the dark.

TOOTHY GRILLE

The mammoth-tooth grille dominates the front aspect of the DeSoto but would be scaled down for 1951. 1950 models are easily spotted by their body-colour vertical grille divider, unique to this year.

SHARED UNIT
All '50 DeSotos shared the same lacklustre straight-six.

SPECIFICATIONS

MODEL DeSoto Custom Convertible (1950)
PRODUCTION 2,900 (1950)
BODY STYLE Two-door convertible.
CONSTRUCTION Steel body and box-section chassis.
ENGINE 236.7cid straight-six.
POWER OUTPUT 112 bhp.
TRANSMISSION Fluid drive semi-automatic.
SUSPENSION *Front:* independent coil springs; *Rear:* leaf springs with live axle.
BRAKES Front and rear drums.
MAXIMUM SPEED 145 km/h (90 mph)
0–60 MPH (0–96 KM/H) 22.1 sec
A.F.C. 6.4 km/l (18 mpg)

ENGINE
The side-valve straight-six was stodgy, putting out a modest 112 bhp.

ADVERT
During the 1950s, car advertising copy became extravagant, relying more on hyperbole than fact. This DeSoto promotion was no exception.

DE TOMASO *Pantera GT5*

AN UNCOMPLICATED SUPERCAR, the Pantera was a charming amalgam of Detroit grunt and Italian glam. Launched in 1971 and sold in North America by Ford's Lincoln-Mercury dealers, it was powered by a mid-mounted Ford 5.7-litre V8 that could muster 256 km/h (159 mph) and belt to 60 mph (96 km/h) in under six seconds. The formidable 350 bhp GT5 was built after Ford pulled out in 1974 and De Tomaso merged with Maserati. With a propensity for the front lifting at speed, hopeless rear visibility, no headroom, awkward seats, and impossibly placed pedals, the Pantera is massively flawed, yet remarkably easy to drive. Handling is poised and accurate, plus that wall of power which catapults the car to 48 km/h (30 mph) in less time than it takes to pronounce its name.

BOOT
Lift-up rear panel gave total engine accessibility for maintenance.

HOT BLOCK
Early Panteras would overheat and owners would often see the temperature gauge creep past 110°C (230°F).

ALL SHOOK UP
Elvis Presley shot his Pantera when it wouldn't start.

EXHAUSTS
Four exhausts were necessary to provide an efficient outlet for all that power.

LIMITED HEADROOM
Do not buy a Pantera if you are over 178 cm (5 ft 10 in) tall – there is no headroom.

CONSTRUCTION
The underside was old-fashioned welded pressed steel monocoque.

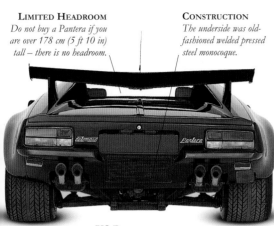

SPECIFICATIONS

MODEL De Tomaso Pantera GT5 (1974–93)

PRODUCTION N/A

BODY STYLE Mid-engined two-seater coupé.

CONSTRUCTION Pressed-steel chassis body unit.

ENGINE 5763cc V8.

POWER OUTPUT 350 bhp at 6000 rpm.

TRANSMISSION Five-speed manual ZF Transaxle.

SUSPENSION All-round independent.

BRAKES All-round ventilated discs.

MAXIMUM SPEED 256 km/h (159 mph)

0–60 MPH (0–96 KM/H) 5.5 sec

0–100 MPH (0–161 KM/H) 13.5 sec

A.F.C. 5.3 km/l (15 mpg)

US RESTRICTIONS
Americans were not able to buy the proper GT5 due to the car's lack of engine-emission controls and had to settle for just the GT5 badges.

STYLING
Shape was penned by Tom Tjaarda, who gave it a clean uncluttered nose.

STUNNING PROFILE
Fat arches, aggressive GT5 graphics down the flanks, 28-cm (11-in) wide wheels, and ground clearance you could not slide an envelope under make the Pantera look evil.

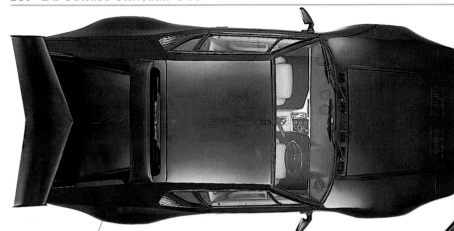

WHEELARCH

Wheelarches strained outwards to cover 33-cm (13-in) rear tyres.

PANTERA AT SPEED

The huge wing helps rear down-force but actually slows the Pantera down. At the General Motors Millbrook proving ground in England, a GT5 with the wing in place made 238 km/h (148 mph); without the wing it reached 244 km/h (151.7 mph).

COCKPIT

With the engine so close to the interior, the cabin temperature could get very hot.

INTERIOR

The Pantera requires a typical Italian driving position – long arms and short legs. Switches and dials are all over the place, but the glorious engine tone is right next to your ears.

TYRES

Giant Pirelli P7 345/45 rear rubber belonged on the track and gave astonishing road traction.

TRANSAXLE
The ZF transaxle was also used in the Ford GT40 (see pages 258–61) and cost more to make than the engine.

SHARED ENGINEERING
The Pantera was engineered by Giampaolo Dallara, also responsible for the Lamborghini Miura (see pages 318–21).

ENGINE
The Pantera is really just a big power plant with a body attached. The monster V8 lives in the middle, mated to a beautifully built aluminium-cased ZF transaxle.

FRONT-END SCARES
Despite a front spoiler, the little weight upfront meant that when the Pantera hit over 193 km/h (120 mph), the nose would lift and the steering would lighten up alarmingly. Generally, though, the car's rear-wheel drive set-up made for neat, controllable handling; an expert could literally steer the Pantera on the throttle.

DODGE *Charger R/T*

COLLECTORS RANK THE 1968 Dodge Charger as one of the fastest and best-styled muscle cars of its era. This, the second generation of Charger, marked the pinnacle of the horsepower race between American car manufacturers in the late 1960s. At that time, gasoline was 10 cents a gallon, Americans had more disposable income than ever before, and engine capacity was everything to the aspiring car buyer. With its hugely powerful 7.2-litre engine, the Charger 440 was, in reality, a thinly veiled street racer. The Rapid Transit (R/T) version was a high-performance factory option, which included heavy-duty suspension and brakes, dual exhausts, and wider tyres. At idle, the engine produced such massive torque that it rocked the car body from side to side. Buyers took the second generation Charger to their hearts in a big way, with sales outstripping the earlier lacklustre model by a factor of six.

HANDSOME BEAST
The Charger was the creation of Dodge's chief of design, Bill Brownlie, and its clean, voluptuous lines gave this car one of the most handsome shapes of the day. It left you in no doubt as to what it was all about: guts and purpose. The mean-looking nose, blacked-out grille, and low bonnet made drivers of lesser machines move over fast.

INDICATORS
Neat styling features included indicator repeaters built into the bonnet scoop.

ENGINE
The potent engine had enough power to spin the rear wheels in every gear.

WOODEN WHEEL
Factory options included wood-grained steering wheel and cruise control.

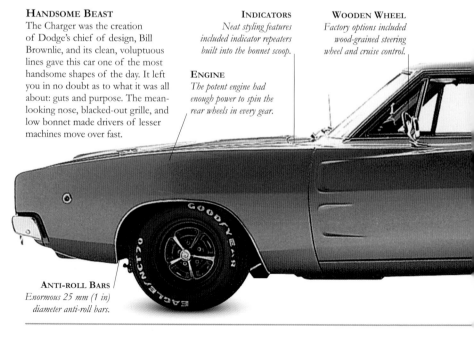

ANTI-ROLL BARS
Enormous 25 mm (1 in) diameter anti-roll bars.

REAR STYLING

"Buttress-backed" styling was America's version of a European 2+2 sports coupé. Ads called the Charger "A beautiful screamer", which was aimed at "a rugged type of individual". Profile is all-agression, with lantern-jawed lines, mock vents on the doors, bumble bee stripes and twin exhausts that roared.

SPECIFICATIONS

MODEL Dodge Charger (1967–70)

PRODUCTION 96,100

BODY STYLE Two-door, four-seater.

CONSTRUCTION Steel monocoque body.

ENGINE 7.2-litre V8.

POWER OUTPUT 375 bhp at 3200 rpm.

TRANSMISSION Three-speed TorqueFlite auto, or Hurst four-speed manual.

SUSPENSION *Front:* heavy duty independent; *Rear:* leaf-spring.

BRAKES Heavy duty, 280 mm (11 in) drums, with optional front discs.

MAXIMUM SPEED 241 km/h (150 mph)

0–60 MPH (0–96 KM/H) 6 sec

0–100 MPH (0–161 KM/H) 13.3 sec

A.F.C. 3.5 km/l (10 mpg)

SEATS
Bucket seats were de rigueur *at the time.*

"SOFT" INTERIOR
Chargers were also for those "who like it soft inside". All had standard clock, heater, and cigarette lighter.

SECURITY
The chrome, quick-fill, racing-style gas cap was attached to the car by wire to stop souvenir-hunters.

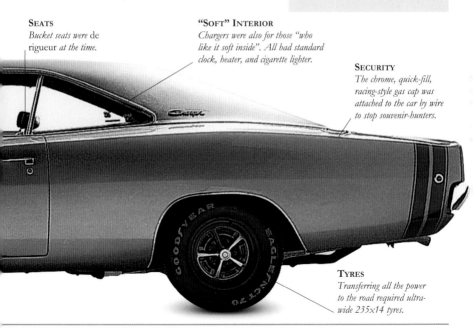

TYRES
Transferring all the power to the road required ultra-wide 235x14 tyres.

STEERING WHEEL
*Huge steering wheel was
essential for keeping all
that grunt in a straight line.*

INTERIOR
The standard R/T cockpit is
functional to the point of being
stark. No distractions here –
just a matt black dash with six
gauges that included a 150 mph
(241 km/h) speedometer.

COLOURS
*Choices originally
included Plum
Crazy, Go Mango,
and Top Banana.*

FUEL
*The gargantuan
engine returned
just 3.5 km/l
(10 mpg).*

LIGHTS
*Hazard warning
lights were groovy
features for 1967.*

NEVAD

OC

STAR OF THE SCREEN

A car with star quality, the Charger featured in the classic nine-minute chase sequence in the film *Bullitt*. It also had major roles in the 1970s cult movie *Vanishing Point*, and the American television series, *The Dukes of Hazzard*.

ENGINE

The wall-to-wall engine found in the R/T Charger is Dodge's immensely powerful 440 Magnum – a 7.2-litre V8. This stump-pulling power plant produced maximum torque at a lazy 3200 rpm – making it obscenely quick, yet as docile as a kitten in town traffic.

HEADLIGHTS

These were hidden under electric flaps to give the Charger a sinister grin.

EDSEL *Bermuda*

WITHOUT THAT INFAMOUS GRILLE, the Bermuda wouldn't have been a bad old barge. The rest looked pretty safe and suburban, and even those faddish rear lights weren't that offensive. At $3,155 it was the top Edsel wagon, wooing the WASPs with more mock wood than Disneyland. But Ford had oversold the Edsel big-time, and every model suffered guilt by association. Initial sales in 1957 were nothing like the predicted 200,000, but weren't disastrous either. The Bermudas, though, found just 2,235 buyers and were discontinued after only one year. By '58, people no longer believed the hype, and Edsel sales evaporated; the company ceased trading in November 1959. Everybody knew that the '58 recession killed the Edsel, but at Ford major players in the project were cruelly demoted or fired.

ODD STYLING

Looking back, one wonders how one of the most powerful corporations in the world could possibly have signed off on such a stylistic debacle. '58 Edsels weren't just ugly, they were appallingly weird. The Bermuda's side view, however, is innocuous enough and no worse than many half-timbered shopping-mall wagons of the period.

AERIAL
Push-button radio with manual antenna was an expensive $95 option.

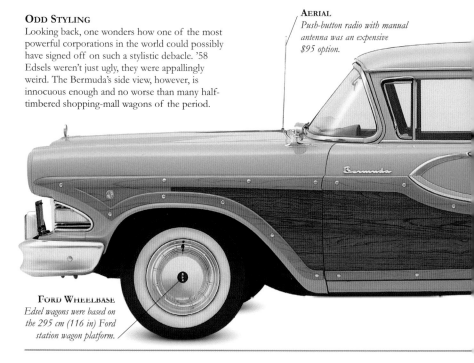

FORD WHEELBASE
Edsel wagons were based on the 295 cm (116 in) Ford station wagon platform.

FRONT ASPECT

The grille was so prominent that it required separate flanking bumpers. The Edsel mascot adorns the front of the bonnet; the name was chosen from 6,000 possibilities, including Mongoose, Turcotinga, and Utopian Turtletop.

STEERING

49 per cent of all Edsels had power steering.

COLOUR CHOICE

This Bermuda is painted in Spring Green, but buyers had a choice of 161 different colour combinations.

ROOF KINK

Note how the roof is slightly kinked to give the huge panel extra rigidity.

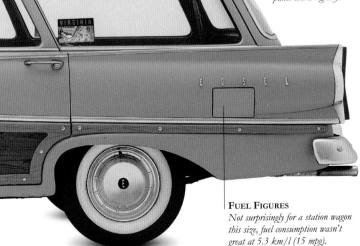

FUEL FIGURES

Not surprisingly for a station wagon this size, fuel consumption wasn't great at 5.3 km/l (15 mpg).

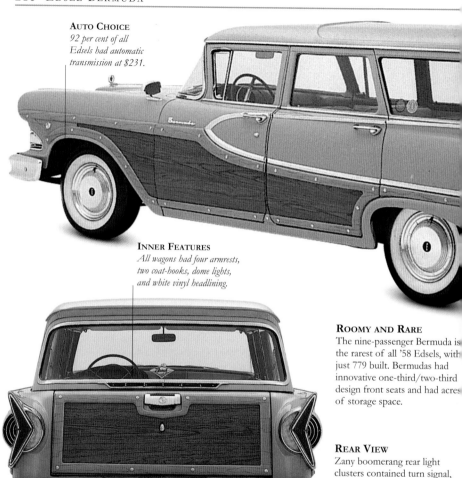

AUTO CHOICE
92 per cent of all Edsels had automatic transmission at $231.

INNER FEATURES
All wagons had four armrests, two coat-hooks, dome lights, and white vinyl headlining.

ROOMY AND RARE
The nine-passenger Bermuda is the rarest of all '58 Edsels, with just 779 built. Bermudas had innovative one-third/two-third design front seats and had acres of storage space.

REAR VIEW
Zany boomerang rear light clusters contained turn signal, stop, and back-up lights. Despite later criticism of the models' design, advance publicity ensured that 4,000 Edsels were sold when they were launched on "Edsel Day", 4 September 1957.

SPECIFICATIONS

MODEL Edsel Bermuda (1958)

PRODUCTION 1,456 (1958, six-seater Bermudas)

BODY STYLE Four-door, six-seater station wagon.

CONSTRUCTION Steel body and chassis.

ENGINE 361cid V8.

POWER OUTPUT 303 bhp.

TRANSMISSION Three-speed manual with optional overdrive, optional three-speed automatic with or without Teletouch control.

SUSPENSION *Front:* independent coil springs; *Rear:* leaf springs with live axle.

BRAKES Front and rear drums.

MAXIMUM SPEED 174 km/h (108 mph)

0–60 MPH (0–96 KM/H) 10.2 sec

A.F.C. 5.3 km/l (15 mpg)

SUSPENSION
Rear suspension was by leaf springs.

TELETOUCH
Teletouch button sent a signal to the car's "precision brain".

ENGINE

"They're the industry's newest – and the best", cried the advertising. Edsel engines were strong 361 or 410cid V8s, with the station wagons usually powered by the smaller unit. The E400 on the valve covers indicates the unit's amount of torque.

INTERIOR

Never one of Edsel's strongest selling points, the Teletouch gear selector was operated by push-buttons in the centre of the steering wheel. It was gimmicky and unreliable.

EDSEL *Corsair*

BY 1959 AMERICA HAD LOST HER confidence; the economy nose-dived, Russia was first in space, there were race riots in Little Rock, and Ford was counting the cost of their disastrous Edsel project – close on 400 million dollars. "The Edsel look is here to stay" brayed the adverts, but the bold new vertical grille had become a country-wide joke. Sales didn't just die, they never took off, and those who had been rash enough to buy hid their chromium follies in suburban garages. Eisenhower's mantra of materialism was over, and buyers wanted to know more about economical compacts like the Nash Rambler, Studebaker Lark, and novel VW Beetle. Throw in a confusing 18-model line-up, poor build quality, and disenchanted dealers, and "The Newest Thing on Wheels" never stood a chance. Now famous as a powerful symbol of failure, the Edsel stands as a telling memorial to the foolishness of consumer culture in Fifties America.

A RE-HASHED FORD

By 1959, the Corsair had become just a restyled Ranger, based on the Ford Fairlane. Corsairs had bigger motors and more standard equipment. But even a sticker price of $3,000 for the convertible didn't help sales, which were a miserable model year total of 45,000. Ford were desperate and tried to sell it as "A new kind of car that makes sense".

WING MIRROR
The hooded chrome door mirror was remote-controlled, an extremely rare after-market option.

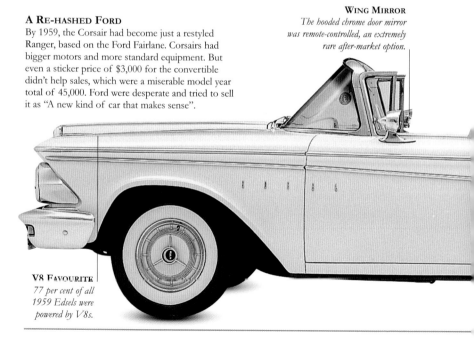

V8 FAVOURITE
77 per cent of all 1959 Edsels were powered by V8s.

SPECIFICATIONS

MODEL Edsel Corsair Convertible (1959)

PRODUCTION 1,343 (1959)

BODY STYLE Four-seater coupé.

CONSTRUCTION Steel body and chassis.

ENGINES 332cid, 361cid V8s.

POWER OUTPUT 225–303 bhp.

TRANSMISSION Three-speed manual with optional overdrive, optional two- or three-speed Mile-O-Matic automatic.

SUSPENSION *Front:* independent with coil springs; *Rear:* leaf springs with live axle.

BRAKES Front and rear drums.

MAXIMUM SPEED 153–169 km/h (95–105 mph)

0–60 MPH (0–96 KM/H) 11–16 sec

A.F.C. 5.3 km/l (15 mpg)

EMPTY ADVERTISING
Ford's Edsel arrived in 1957 on the back of intense TV and magazine coverage. But by the time it hit the showrooms, the market had done a *volte-face* and wanted more than just empty chromium rhetoric.

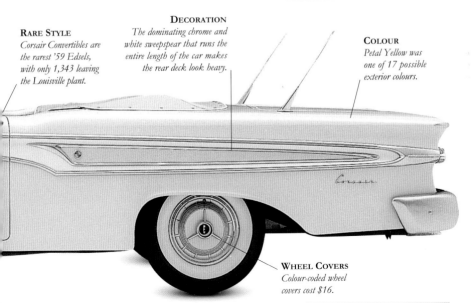

RARE STYLE
Corsair Convertibles are the rarest '59 Edsels, with only 1,343 leaving the Louisville plant.

DECORATION
The dominating chrome and white sweepspear that runs the entire length of the car makes the rear deck look heavy.

COLOUR
Petal Yellow was one of 17 possible exterior colours.

WHEEL COVERS
Colour-coded wheel covers cost $16.

TOILET SEAT STYLING
Roy Brown, the Edsel's designer,
claimed that "The front theme
of our newest car combines
nostalgia with modern vertical
thrust". Other pundits were
not so positive and
compared it to a
horse collar, a man
sucking a lemon, or
even a toilet seat.

WEIGHT
*Weighing in at a
considerable 1,71
kg (3,790 lb) th
convertible was
heavier than
the sedan.*

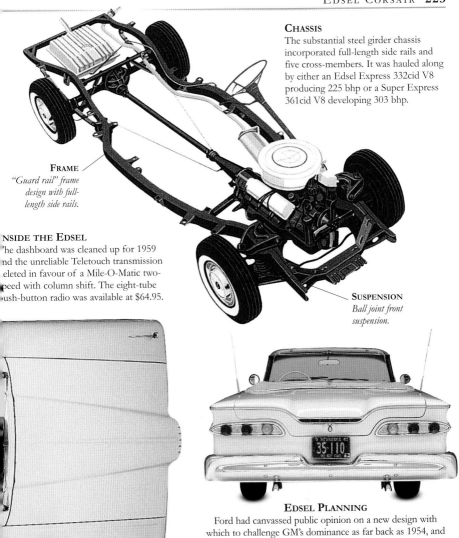

CHASSIS
The substantial steel girder chassis incorporated full-length side rails and five cross-members. It was hauled along by either an Edsel Express 332cid V8 producing 225 bhp or a Super Express 361cid V8 developing 303 bhp.

FRAME
"Guard rail" frame design with full-length side rails.

INSIDE THE EDSEL
The dashboard was cleaned up for 1959 and the unreliable Teletouch transmission deleted in favour of a Mile-O-Matic two-speed with column shift. The eight-tube push-button radio was available at $64.95.

SUSPENSION
Ball joint front suspension.

EDSEL PLANNING
Ford had canvassed public opinion on a new design with which to challenge GM's dominance as far back as 1954, and named the new project the E ("experimental") Car. By the time it appeared, it was a ridiculous leviathan.

FACEL *Vega II*

WHEN SOMEONE LIKE PABLO PICASSO chooses a car, it is going to look good. In its day, the Facel II was a poem in steel and easily as beautiful as anything turned out by the Italian styling houses. Small wonder then that Facels were synonymous with the Sixties' jet set. Driven by Ringo Starr, Ava Gardner, Danny Kaye, Tony Curtis, François Truffaut, and Joan Fontaine, Facels were one of the most charismatic cars of the day. Even death gave them glamour; the novelist Albert Camus died while being passengered in his publisher's FVS in January 1960. In 1961, the HK 500 was reskinned and given cleaner lines, an extra 15 cm (6 in) in length, and dubbed the Facel II. At 1.5 tonnes, the II was lighter than the 500 and could storm to 225 km/h (140 mph). Costing more than the contemporary Aston Martin DB4 *(see pages 32–35)* and Maserati 3500, the Facel II was as immortal as a Duesenberg, Hispano Suiza, or Delahaye. We will never see its like again.

HAND-CRAFTED SUPERCAR
In terms of finish, image, and quality, Facel Vegas were one of the most successful hand-made supercars. Body joints were perfectly flush, doors closed like heavy vaults, brightwork was stainless steel, and even the roof line was fabricated from five seamlessly joined sections.

REAR SEATING
The leather back seat folded down to make a luggage platform.

BUMPER
Bumper is not chrome but rust-resistant stainless steel.

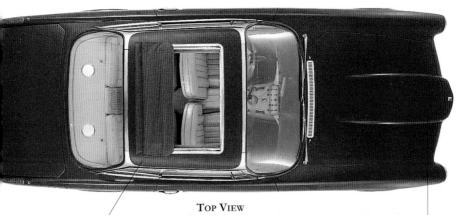

SUNROOF
Fabric, roll-back, full-length sunroof was a period after-market accessory.

TOP VIEW
Facel II used the same wheelbase and engine as the HK 500, but the shape was refined to make it look more modern, losing such cliches as the dated wrap-around windscreen.

POWER BULGE
Prodigious bonnet bulge cleared air cleaners and twin carbs.

GEARBOX
Manual Pont-a-Mousson gearbox began life in a truck.

FUEL CONSUMPTION
Driven fast, the Facel II would drink one gallon of fuel every ten miles.

SPINNERS
Knock-off wheel spinners.

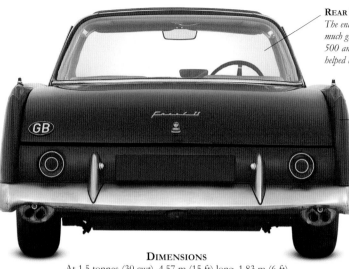

REAR VISIBILITY
The enlarged rear window gave a much greater glass area than the HK 500 and almost 90 per cent visibility helped by slimmer screen pillars.

GENERAL MANUFACTURERS
In the '50s, Facel made motor scooters, jet engines, office furniture, and kitchen cabinets.

DIMENSIONS
At 1.5 tonnes (30 cwt), 4.57 m (15 ft) long, 1.83 m (6 ft) wide, and only 1.3 m (4 ft 3 in) high, the Facel II aped the girth and bulk of contemporary American iron.

SUSPENSION
Selectaride shock absorbers provided a comfortable ride.

DOMINATING GRILLE
The intimidating frontage is all grille because the hot-running V8 engine needed all the cooling air it could get. HK 500 had four round headlamps, but the Facel II's voguish stacked lights were shamelessly culled from contemporary Mercedes saloons.

INTERIOR
Steering wheel points straight to the driver's heart. Note the unmistakable aircraft-type panel layout with centre gauges and heater controls like hand throttles.

MOOTH LIGHTING
rake-indicator lights are cut out
f the rear wings and help to
nhance the Facel's seamless lines.
o achieve this stunning one-
iece look, the car's light alloy
ody panels were hand-finished
nd mated to each other.

BODY STYLING
*Rakish body was
artistically similar to
the Facellia Coupé.*

RARE MOTOR
*By far the rarest Facel with only
184 made, IIs are still fiercely
admired by Facel fanciers.*

SPECIFICATIONS

MODEL Facel Vega Facel II (1962–64)

PRODUCTION 184

BODY STYLE Two-door, four-seater
Grand Tourer.

CONSTRUCTION Steel chassis, steel/light
alloy body.

ENGINE 6286cc cast-iron V8.

POWER OUTPUT 390 bhp at 5400 rpm
(manual), 355 bhp at 4800 rpm (auto).

TRANSMISSION Three-speed TorqueFlite
auto or four-speed Pont-a-Mousson manual.

SUSPENSION Independent front coil
springs, rear live axle leaf springs.

BRAKES Four-wheel Dunlop discs.

MAXIMUM SPEED 240 km/h (149 mph)

0–60 MPH (0–96 KM/H) 8.3 sec

0–100 MPH (0–161 KM/H) 17.0 sec

A.F.C. 5.4 km/l (15 mpg)

BONNET
*Bonnet lid was
huge, but then so
was the engine.*

BRAKES
*Disc brakes all-
round countered
the Facel's
immense power.*

FERRARI *250 GT SWB*

IN AN ERA WHEN FERRARI WAS turning out some lacklustre road cars, the 250 GT SWB became a yardstick, the car against which all other GTs were judged and one of the finest Ferraris ever. Of the 167 made between 1959 and 1962, 74 were competition cars – their simplicity made them one of the most competitive sports racers of the Fifties. Built around a tubular chassis, the V12 3.0 engine lives at the front, along with a simple four-speed gearbox with Porsche internals. But it is that delectable Pininfarina-sculpted shape that is so special. Tense, urgent, but friendly, those smooth lines have none of the intimidating presence of a Testarossa or Daytona. The SWB stands alone as a perfect blend of form and function – one of the world's prettiest cars, and on the track one of the most successful. The SWB won races from Spa to Le Mans, Nassau to the Nürburgring. Which is exactly what Enzo Ferrari wanted. "They are cars", he said, "which the sporting client can use on the road during the week and race on Sundays". Happy days.

DESIGN CREDITS
Soft, compact, and rounded, Pininfarina executed the design, while Scaglietti took care of the sheet metal. The result was one of the most charismatic cars ever produced.

NO CLEANERS
Instead of air cleaners, competition cars used filterless air trumpets.

AIR SCOOPS
Wing air scoops helped to cool the engine.

NOSE
Gently tapering nose is a masterpiece of the panel-beater's art.

ENGINE

The V12 power unit had a seven-bearing crankshaft turned from a solid billet of steel, single plug per cylinder, and three twin-choke Weber DCL3 or DCL6 carburettors.

SPECIFICATIONS

MODEL Ferrari 250 GT SWB (1959–62)

PRODUCTION 167 (10 RHD)

BODY STYLE Two-seater GT coupé.

CONSTRUCTION Tubular chassis with all-alloy or alloy/steel body.

ENGINE 2953cc V12.

POWER OUTPUT 280 bhp at 7000 rpm.

TRANSMISSION Four-speed manual.

SUSPENSION Independent front coil and wishbones, rear live axle leaf springs.

BRAKES Four-wheel discs.

MAXIMUM SPEED 237 km/h (147 mph)

0–60 MPH (0–96 KM/H) 6.6 sec

0–100 MPH (0–161 KM/H) 16.2 sec

A.F.C. 4.2 km/l (12 mpg)

REAR SCREEN

Expansive rear window sat above enormous 123-litre (27-gallon) fuel tank.

WHEELS

The SWB sat on elegant, chrome-plated Borrani competition wire wheels.

OVERHEAD VIEW

The car has perfect balance. Shape is rounded and fluid and the first 11 SWBs were built in alloy, though these rare lightweight models suffered from stretching alloy. Road cars had a steel body and aluminium bonnet and doors.

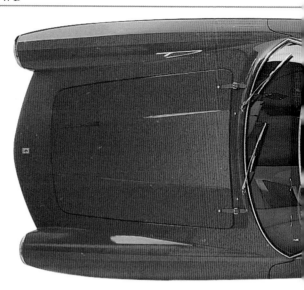

STRAP 'EM IN

The 250's roll-cage and modern harnesses were sops to safety, but understandable considering that progressively more power was extracted from the V12 engine.

UNDERSTATED BEAUTY

The 250 GT is a polished gem, hugging the road limpet-low. Front combines beauty and threat with steely grin and squat wheelarch-filling attitude. Nothing is exaggerated for effect.

ROAD PROTECTION

Unlike this race car, ro *cars had vestigial front* *bumpers and the pranci* *horse badge in the grille.*

INTERIOR
Despite the matinée idol exterior, the interior is a place of work. Functional fascia is basic crackle black with no frills. Sun visors were notably absent. The cockpit was snug and airy but noisy when the key was turned.

FILLER CAP
Huge alloy filler-cap was to allow fast petrol stops.

RACING STATEMENT
Two sets of aggressive drainpipe twin exhausts dominate the SWB's rump and declare its competition bloodline. For many years the 250 GT dominated hill climbs and track meets all over the world. The SWB 250 GT was the ultimate racer.

FERRARI *275 GTB/4*

THE GTB/4 was a hybrid made for two short years from 1966 to 1968. With just 350 built, a mere 27 in right-hand drive, it was not one of Ferrari's money-spinners. So named for its four camshafts, the GTB still ranks as the finest road car Ferrari produced before Fiat took control of the company. With fully independent suspension, a five-speed gearbox, and a fetching Pininfarina-designed and Scaglietti-built body, it was the last of the proper Berlinettas. Nimble and compact, with neutral handling and stunning design, this is probably one of the most desirable Ferraris ever made.

SPECIFICATIONS

MODEL Ferrari 275 GTB/4 (1966–68)
PRODUCTION 350
BODY STYLE Two-seater front-engined coupé.
CONSTRUCTION Steel chassis, aluminium body.
ENGINE 3.3-litre twin overhead-cam dry sump V12.
POWER OUTPUT 300 bhp at 8000 rpm.
TRANSMISSION Five-speed all-synchromesh.
SUSPENSION All-round independent.
BRAKES Four-wheel servo discs.
MAXIMUM SPEED 257 km/h (160 mph)
0–60 MPH (0–96 KM/H) 5.5 sec
0–100 MPH (0–161 KM/H) 13 sec
A.F.C. 4.2 km/l (12 mpg)

THE MECHANICS
This was Ferrari's first ever production four-cam V12 engine and their first road-going prancing horse with an independent rear end. The type 226 engine was related to the 330 P2 prototypes of the 1965 racing season. The GTB/4's chassis is made up of a ladder frame built around two oval-tube members.

A MOTORING BEAUTY
The GTB/4 is prettier than an E-Type *(see pages 306–09)*, Aston Martin DB4 *(see pages 32–35)*, or Lamborghini Miura *(see pages 318–21)*. The small boot, small cockpit, and long nose are classic Pininfarina styling – an arresting amalgam of beauty and brawn. The interior, though, is trimmed in unluxurious vinyl.

FERRARI *Daytona*

THE CLASSICALLY sculptured and outrageously
quick Daytona was a supercar with a split personality.
Under 193 km/h (120 mph), it felt like a truck with
heavy inert controls and crashing suspension. But
once the needle was heading for 225 km/h (140
mph), things started to sparkle. With a romantic flat-
out maximum of 280 km/h (170 mph), it was the last
of the great front-engined V12 war horses. Launched
at the 1968 Paris Salon as the 365 GTB/4, the press
immediately named it "Daytona" in honour of
Ferrari's success at the 1967 American 24-hour race.
Faster than all its Italian and British contemporaries,
the chisel-nosed Ferrari won laurels on the race-track
as well as the hearts and pockets of wealthy
enthusiasts all over the world.

POEM IN STEEL
A poem in steel, only a handful of
other cars could be considered in the
same aesthetic league as the Daytona.

INSIDE AND OUT
With hammock-type racing seats,
a cornucopia of black-on-white
instruments, and a provocatively
angled, extra-long gear shift, the
cabin promises some serious
excitement. Beneath the exterior is
a skeleton of chrome-molybdenum
tube members, giving enormous
rigidity and strength.

SPECIFICATIONS

MODEL Ferrari 365 GTB/4 Daytona (1968–73)

PRODUCTION 1,426 (165 RHD models)

BODY STYLE Two-seater fastback.

CONSTRUCTION Steel/alloy/glass-fibre body, separate multi-tube chassis frame.

ENGINE V12 4390cc.

POWER OUTPUT 352 bhp at 7500 rpm.

TRANSMISSION Five-speed all-synchromesh.

SUSPENSION Independent front and rear.

BRAKES Four-wheel discs.

MAXIMUM SPEED 280 km/h (174 mph)

0–60 MPH (0–96 KM/H) 5.4 sec

0–100 MPH (0–161 KM/H) 12.8 sec

A.F.C. 5 km/l (14 mpg)

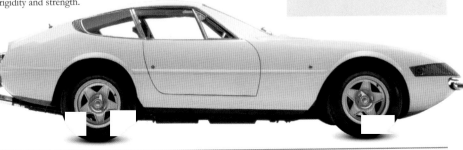

FERRARI *Dino 246 GT*

PRETTY ENOUGH TO STOP a speeding train, the Dino came not from Enzo Ferrari's head, but from his heart. The Dino was a tribute to the great man's love for his son, Alfredino, who died of a kidney disease. Aimed at the Porsche 911 buyer *(see pages 420–21)*, the 246 Dino engine came with only half the number of cylinders usually found in a Ferrari. Instead of a V12 configuration, it boasted a 2.4-litre V6 engine, yet was nonetheless capable of a very Ferrari-like 241 km/h (150 mph). With sparkling performance, small girth, and mid-engined layout, it handled like a go-kart, and could be hustled around with enormous aplomb. Beautifully sculpted by Pininfarina, the 246 won worldwide acclaim as the high point of 1970s automotive styling. In its day, it was among the most fashionable cars money could buy. The rarest Dino is the GTS, with Targa detachable roof panel. The Dino's finest hour was when it was driven by Tony Curtis in the Seventies' ITC television series *The Persuaders*.

BODY CONSTRUCTION

Early Dinos were constructed from alloy, later ones from steel, with the bodies built by Italian designer Scaglietti. Unfortunately, little attention was paid to rust protection. Vulnerable interior body joints and cavities were covered with only a very thin coat of paint.

AERODYNAMICS

The sleek aerodynamic shape of the roof line helped to give the car its impressively high top speed.

TYRES

Wide tyres were essential to deliver the Dino's lithe handling.

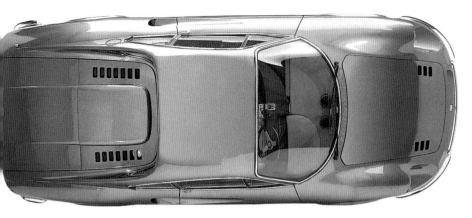

WINDSCREEN
Windscreens do not come much more steeply raked than this one.

REAR ENGINE
The transversely mounted 2418cc V6 has four overhead-cams, a four-bearing crankshaft, and breathes through three twin-choke Weber 40 DCF carburettors. Power output is 195 bhp. This particular engine's distinctive throaty roar is a Ferrari legend.

DINO PRICES
Prices went crazy in the Eighties, but are now half that value.

ENGINE POSITION

The engine is positioned in the middle of the car, which gives mechanics little space to work in. The spare wheel and battery are located under the bonnet in the front, leaving very little room to carry extras such as luggage. Optional perspex headlamp cowls can increase the Dino's top speed by 5 km/h (3 mph).

BADGING

246s wore the Dino badge on the nose, never the Ferrari's prancing horse.

SPECIFICATIONS

MODEL Ferrari Dino 246 GT (1969–74)
PRODUCTION 2,487
BODY STYLE Two-door, two seater.
CONSTRUCTION Steel body, tubular frame.
ENGINE Transverse V6/2.4 litre.
POWER OUTPUT 195 bhp at 5000 rpm.
TRANSMISSION Five-speed, all-synchromesh.
SUSPENSION Independent front and rear.
BRAKES Ventilated discs all round.
MAXIMUM SPEED 238 km/h (148 mph)
0–60 MPH (0–96 KM/H) 7.1 sec
0–100 MPH (0–161 KM/H) 17.6 sec
A.F.C. 7.8 km/l (22 mpg)

CURVY ITALIAN

The sensuous curves are unmistakably supplied by Ferrari. The Ferrari badge and prancing horse were fitted by a later owner. The thin original paint job means that most surviving Dinos will have had at least one body rebuild by now.

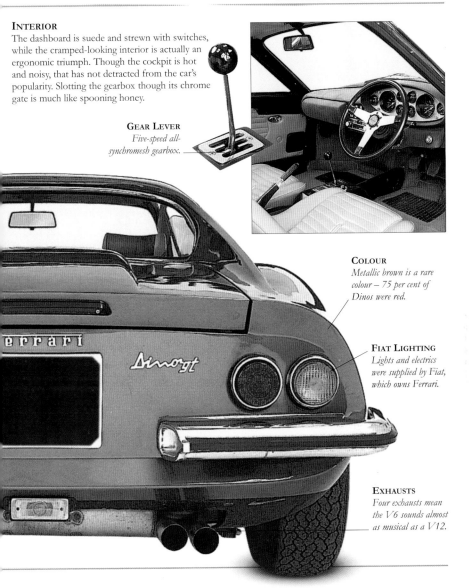

INTERIOR
The dashboard is suede and strewn with switches, while the cramped-looking interior is actually an ergonomic triumph. Though the cockpit is hot and noisy, that has not detracted from the car's popularity. Slotting the gearbox though its chrome gate is much like spooning honey.

GEAR LEVER
Five-speed all-synchromesh gearbox.

COLOUR
Metallic brown is a rare colour – 75 per cent of Dinos were red.

FIAT LIGHTING
Lights and electrics were supplied by Fiat, which owns Ferrari.

EXHAUSTS
Four exhausts mean the V6 sounds almost as musical as a V12.

Ferrari *365 GT4 Berlinetta Boxer*

The Berlinetta Boxer was meant to be the jewel in Ferrari's crown – one of the fastest GT cars ever. Replacing the legendary V12 Ferrari Daytona *(see page 233)*, the 365 BB was powered by a flat-12 "Boxer" engine, so named for the image of the horizontally located pistons punching at their opposite numbers. Mid-engined, with a tubular chassis frame and clothed in a peerless Pininfarina-designed body (a mixture of alloy, glass-fibre, and steel), the 365 was assembled by Scaglietti in Modena. First unveiled in 1971 at the Turin Motor Show, the formidable 4.4-litre 380 bhp Boxer was so complex that deliveries to buyers did not start until 1973. The trouble was that Ferrari had suggested that the Boxer could top 298 km/h (185 mph), when it could actually only manage around 274 km/h (170 mph), slightly slower than the outgoing Daytona. In 1976 Ferrari replaced the 365 with the five-litre Boxer 512, yet the 365 is the faster and rarer model, with only 387 built.

Classic Money

In the classic car boom of the mid-Eighties, Boxers changed hands for mad money. The 512 trebled in value before the crash, with the 365 doubling its price. Now both machines have fallen back to realistic levels.

Fuel Capacity

The Boxer could carry 120 litres (26 gallons) of petrol.

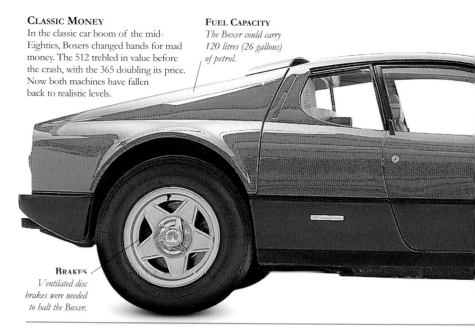

Brakes
Ventilated disc brakes were needed to halt the Boxer.

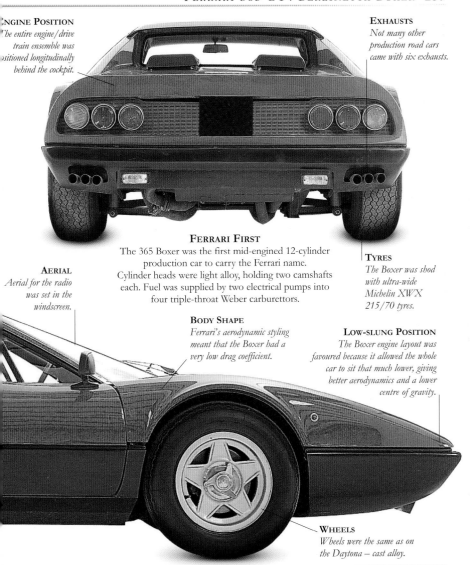

ENGINE POSITION
The entire engine/drive train ensemble was positioned longitudinally behind the cockpit.

EXHAUSTS
Not many other production road cars came with six exhausts.

FERRARI FIRST
The 365 Boxer was the first mid-engined 12-cylinder production car to carry the Ferrari name. Cylinder heads were light alloy, holding two camshafts each. Fuel was supplied by two electrical pumps into four triple-throat Weber carburettors.

AERIAL
Aerial for the radio was set in the windscreen.

TYRES
The Boxer was shod with ultra-wide Michelin XWX 215/70 tyres.

BODY SHAPE
Ferrari's aerodynamic styling meant that the Boxer had a very low drag coefficient.

LOW-SLUNG POSITION
The Boxer engine layout was favoured because it allowed the whole car to sit that much lower, giving better aerodynamics and a lower centre of gravity.

WHEELS
Wheels were the same as on the Daytona – cast alloy.

PROTOTYPE TESTING
A handful of Boxer prototypes were subject to extensive testing. Pre-production cars were recognizable by a number of differences, one being the roof-mounted radio aerial – factory cars had them enclosed in the windscreen. Pininfarina's shape went virtually unchanged from the prototype into the production version.

INTERIOR
An amalgam of racer and grand tourer, the Boxer's cabin was functional yet luxurious, with electric windows and air-conditioning. Switches for these were positioned on the console beneath the gear lever.

CENTRE CONSOLE
The rear-mounted gearbox meant that only a small transmission tunnel was needed, saving cabin room.

ENGINE

A magnificent piece of foundry art, the flat-12 has a crankshaft machined from a solid billet of chrome-molybdenum steel. Instead of timing chains, the 365 used toothed composite belts, an innovation in 1973.

CYLINDERS
The Boxer had twin oil filters, one for each bank of six cylinders.

SPECIFICATIONS

MODEL Ferrari 365 GT4 Berlinetta Boxer (1973–76)

PRODUCTION 387 (58 RHD models)

BODY STYLE Two-seater sports.

CONSTRUCTION Tubular space-frame chassis.

ENGINE 4.4-litre flat-12.

POWER OUTPUT 380 bhp at 7700 rpm.

TRANSMISSION Five-speed all synchromesh rear-mounted gearbox.

SUSPENSION Independent front and rear.

BRAKES Ventilated front and rear discs.

MAXIMUM SPEED 277 km/h (172 mph)

0–60 MPH (0–96 KM/H) 6.5 sec

0–100 MPH (0–161 KM/H) 15 sec

A.F.C. 4.2 km/l (14 mpg)

COOLING VENT
Slatted bonnet cooling vent helped keep interior cabin temperatures down.

CHASSIS
The Boxer's chassis was derived from the Dino (see pages 234–37), with a frame of steel tubes and doors, bellypan, and nose in aluminium.

LOWER BODYWORK
This was glass-fibre, along with the wheelarch liners and bumpers.

FERRARI *308 GTB*

ONE OF THE best-selling Ferraris ever, the 308 GTB started life with a glass-fibre body designed by Pininfarina and built by Scaglietti. Power was courtesy of the V8 3.0 engine and five-speed gearbox inherited from the 308 GT4. With uptown America as the GTB's target market, federal emission regulations made the GTB clean up its act, evolving into a refined and civilized machine with such hi-tech appurtenances as four valves per cylinder and Bosch fuel-injection. Practical and tractable in traffic, it became the 1980s entry-level Ferrari, supplanting the Porsche 911 *(see pages 420–21)* as the standard issue yuppiemobile.

SPECIFICATIONS

MODEL Ferrari 308 GTB (1975–85)

PRODUCTION 712 (308 GTB glass-fibre); 2,185 (308 GTB steel); 3,219 (GTS).

BODY STYLE Two-door, two-seater sports coupé.

CONSTRUCTION Glass-fibre/steel.

ENGINE Mid-mounted transverse dohc 2926cc V8.

POWER OUTPUT 255 bhp at 7600 rpm.

TRANSMISSION Five-speed manual.

SUSPENSION Independent double wishbones/coil springs all round.

BRAKES Ventilated discs all round.

MAXIMUM SPEED 248 km/h (154 mph)

0–60 MPH (0–96 KM/H) 7.3 sec

0–100 MPH (0–161 KM/H) 19.8 sec

A.F.C. 5.7 km/l (16 mpg)

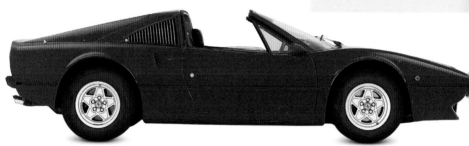

MIXED STYLING CUES
The handsome styling is a blend of Dino 246 and 365 GT4. The Dino provided concave rear windows and conical air intakes, while the 365 brought double bodyshell appearance with a waistline groove. The 2926cc V8 has double overhead cams per bank and four 40 DCNF Weber carburettors.

FRONT ASPECT
With the engine at the back, the wide slatted grille scooped up air for brake and interior ventilation. Retractable, flush-fitting pop-up headlights keep wind force down on the nose and front wheels. The roof on the GTB was always a tin-top; the chic GTS had a Targa top panel.

FERRARI *400 GT*

THE FIRST Ferrari ever offered with automatic transmission, the 400 was aimed at the American market, and was meant to take the prancing horse into the boardrooms of Europe and the US. But the 400's automatic box was a most un-Ferrari-like device, a lazy three-speed GM Turbo-Hydramatic as used in Cadillac, Rolls-Royce, and Jaguar. It may have been the best self-shifter in the world, but it was a radical departure for Maranello, and met with only modest success. The 400 was possibly the most discreet and refined Ferrari ever made. It looked awful in Racing Red – the colour of 70 per cent of Ferraris – so most were finished in dark metallics. The 400 became the 400i GT in 1973 and the 412 in 1985.

SPECIFICATIONS

MODEL Ferrari 400 GT (1976–79)

PRODUCTION 501

BODY STYLE Two-door, four-seater sports saloon.

CONSTRUCTION Steel/alloy body, separate tubular chassis frame.

ENGINE 4390cc twin ohc V12.

POWER OUTPUT 340 bhp at 6800 rpm.

TRANSMISSION Five-speed manual or three-speed automatic.

SUSPENSION Independent double wishbones with coil springs, rear as front with hydro-pneumatic self-levelling.

BRAKES Four-wheel ventilated discs.

MAXIMUM SPEED 241 km/h (150 mph)

0–60 MPH (0–96 KM/H) 7.1 sec

0–100 MPH (0–161 KM/H) 18.7 sec

A.F.C. 4.2 km/l (12 mpg)

365 SIMILARITIES

Apart from the delicate chin spoiler and bolt-on alloys, the shape was pure 365 GT4 2+2. The rectangular design of the body was lightened by a plunging bonnet line and a waist-length indentation running along the 400's flanks.

HEADLIGHTS

Four headlights were retracted into the bodywork by electric motors.

AN-48-TA

FERRARI *Testarossa*

THE TESTAROSSA WAS never one of Modena's best efforts. With its enormous girth and overstuffed appearance, it perfectly sums up the Eighties credo of excess. As soon as it appeared on the world's television screens in *Miami Vice*, the Testarossa, or Redhead, became a symbol of everything that was wrong with a decade of rampant materialism and greed. The Testarossa fell from grace rather suddenly. Dilettante speculators bought them new at £100,000-odd and ballyhooed their values up to a quarter of a million. By 1988, secondhand values were slipping badly and many an investor saw their car shed three-quarters of its value overnight. Today, used Testarossas are highly prized with rising prices and growing investment potential.

RACING LEGEND

Ferrari bestowed on its new creation one of the grandest names from its racing past – the 250 Testa Rossa, of which only 19 were built for retail customers. Design of the new model was determined with the help of Pininfarina's full-sized wind tunnel, but enthusiasts were initially cool about the Testarossa's size and shape.

STYLING

Striking radiator cooling ducts obviated the need to pass water from the front radiator to the mid-mounted engine, freeing the front luggage compartment.

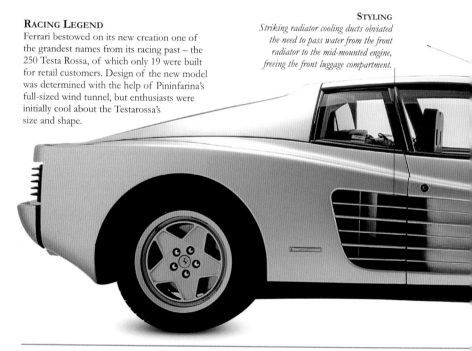

WIDE SUPERCAR

Wider than the Ferrari 512 BB, the Corvette *(see pages 142–45)*, and the Countach *(see pages 322–25)*, it measured a portly 1.83 m (6 ft) across. While this meant a bigger cockpit, the ultra-wide door sills collected mud in wet weather.

SPECIFICATIONS

MODEL Ferrari Testarossa (1988)

PRODUCTION 1,074

BODY STYLE Mid-engined, two-seater sports coupé.

CONSTRUCTION Steel frame with aluminium and glass-fibre panels.

ENGINE Flat-12, 4942cc with dry sump lubrication.

POWER OUTPUT 390 bhp at 6300 rpm.

TRANSMISSION Five-speed manual.

SUSPENSION Independent front and rear.

BRAKES *Front:* disc; *Rear:* drums.

MAXIMUM SPEED 291 km/h (181 mph)

0–60 MPH (0–96 KM/H) 5.3 sec

0–100 MPH (0–161 KM/H) 12.2 sec

A.F.C. 4.2 km/l (12 mpg)

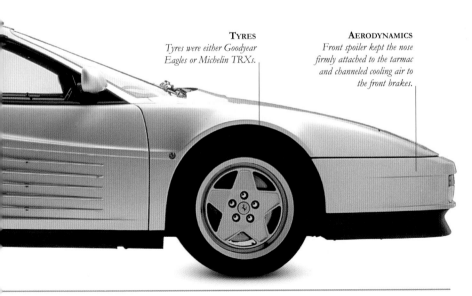

TYRES
Tyres were either Goodyear Eagles or Michelin TRXs.

AERODYNAMICS
Front spoiler kept the nose firmly attached to the tarmac and channeled cooling air to the front brakes.

SPACIOUS INTERIOR
The Testarossa's large body meant plenty of cabin space, with more room for both occupants and luggage. Even so, interior trim was flimsy and looked tired after 112,000 kilometres (70,000 miles).

DOOR MIRRORS
Prominent door mirrors on both sides gave the Testarossa an extra 20 cm (8 in) in width.

REAR-VIEW MIRROR
The curious, periscope-like rear-view mirror was developed by Pininfarina.

COCKPIT
The cockpit was restrained and spartan, with a hand-stitched hide dashboard and little distracting ornamentation. For once a Ferrari's cockpit was accommodating, with electrically adjustable leather seats and air-conditioning as standard.

UAE 576

REAR TREATMENT
Pininfarina's grille treatment was picked up on the rear end, giving stylistic continuity.

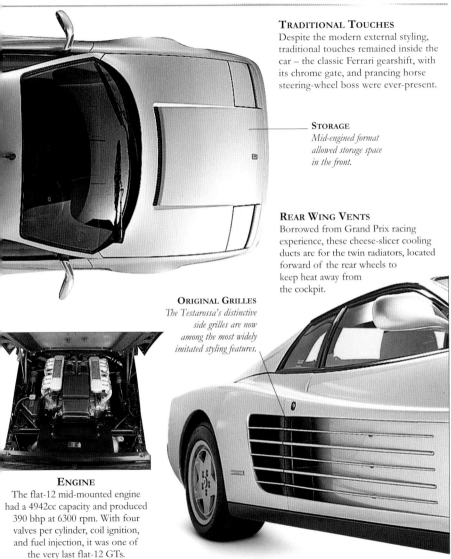

TRADITIONAL TOUCHES
Despite the modern external styling, traditional touches remained inside the car – the classic Ferrari gearshift, with its chrome gate, and prancing horse steering-wheel boss were ever-present.

STORAGE
Mid-engined format allowed storage space in the front.

REAR WING VENTS
Borrowed from Grand Prix racing experience, these cheese-slicer cooling ducts are for the twin radiators, located forward of the rear wheels to keep heat away from the cockpit.

ORIGINAL GRILLES
The Testarossa's distinctive side grilles are now among the most widely imitated styling features.

ENGINE
The flat-12 mid-mounted engine had a 4942cc capacity and produced 390 bhp at 6300 rpm. With four valves per cylinder, coil ignition, and fuel injection, it was one of the very last flat-12 GTs.

FERRARI *456 GT*

A USED FERRARI 456 is one of the world's great supercar bargains. For the price
of a new, hot Ford Focus you can have a beautiful 300 km/h (186 mph) grand
tourer that's also a reliable and practical full four-seater. Strong and capable with
a fine ride and a glorious V12 engine, the 456 is a definite neo-classic. Launched
in 1992 to replace the unloved Mondial, it was the fastest production 2+2 on
the planet and, apart from the F40, the most powerful road car developed by
Ferrari. Slippery and handsome with a carbon-fibre bonnet, pop-up headlamps,
a glorious six-speed gearbox plus an automatic option, the 456 looked and felt
like a Daytona for a fraction of the price. More importantly, the 456 was that
rare thing – a Ferrari with quiet class.

A POEM IN ALLOY
The body was alloy, spot-welded to a steel tubular
chassis using "Feran" filler, and was shaped by extensive
wind tunnel testing. The rear wing was electronically
retractable to give extra down force at speed.
Luggage space was generous, and the
optional special fitted leather
luggage set cost about the
same as a small car.

CLEVER WINDOWS
Electric windows moved down
slightly when you opened the doors.

SPECIFICATIONS

MODEL Ferrari 456 GT (1992)

PRODUCTION 3,289

BODY STYLE Two-door, four-seater coupé.

CONSTRUCTION Alloy panels, tubular steel chassis, composite bonnet.

ENGINE 5,474cc, V12.

POWER OUTPUT 436 bhp.

TRANSMISSION Six-speed transaxle manual.

SUSPENSION Independent all round.

BRAKES Four-wheel ventilated discs.

MAXIMUM SPEED 300 km/h (186 mph)

0–60 MPH (0–96 KM/H) 5 sec

0–100 MPH (0–161 KM/H) 11.6 sec

A.F.C. 6.4 km/l (15 mpg)

BEAUTY BOX

The six-speed transaxle manual gearbox lived at the rear and was an engineering gem, lubricated by its own pump and radiator. GTA had a hydraulic four-speed automatic managed by twin Bosch computers. Chrome gear gate, alloy shifter, and precise action means that manual 456s are worth more than autos.

HIGH LIGHTS

456 was the last Ferrari to have retractable headlights.

ELEGANT INTERIOR

Cabin was high quality and much better than previous Ferraris.

V12 WARHORSE

Brilliant alloy 5.5 litre V12 engine cranked out 436 horsepower.

FERRARI *Enzo*

THE V12 CARBON-FIBRE ENZO is a million-dollar wild child and the most flamboyant Ferrari ever. Good for 364 km/h (226 mph) and capable of 0–100 mph (0–161 km/h) in only 6.6 seconds, the initial production run of 349 units was completely sold out before a single car ever got near a showroom. Ferrari was forced to build another 50 just to please a queue of desperate buyers. Designed by Ken Okuyama of Pininfarina, some say the Enzo is one of the ugliest cars ever, but its show-stopping looks and dramatic doors (similar to the Lamborghini Countach *(see pages 322–25)*) have guaranteed it automotive immortality, and used examples often change hands for more than their original new sticker price. As close as you'll get to a road-going F1 car, the Enzo isn't a supercar or a hypercar – it is best described as the original Wonder Car. The last Enzo ever built, the 400th example, was donated to the Vatican in Rome for charity.

EXPENSIVE EXCESS
The Enzo was the most expensive Ferrari ever built at over £800,000. An oil change cost £600, and if you were unlucky (or careless) enough to exceed the 8,500 rev limit a ruined engine could set you back £150,000 in repairs! Brake pad and tyre replacement could add up to £2,500 per corner.

F1 WHEELS
Wheels are F1 style with single nut.

BASIC ACCOMMODATION
Cabin is stark, simple, and carbon-fibre.

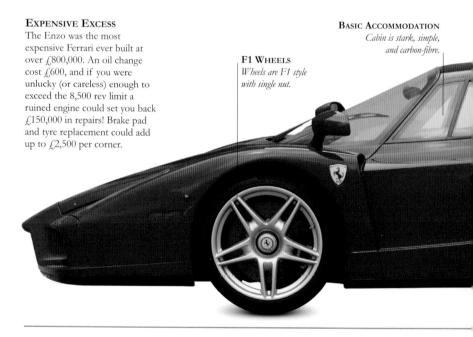

DRAMATIC DOORS
*Enormous doors make
up most of roof area.*

AIR SHOVELS
*Massive nostrils cool
brakes and keep
nose down.*

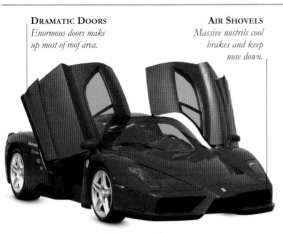

SPECIFICATIONS

MODEL Ferrari Enzo (2002)
PRODUCTION 400
BODY STYLE Two-door, two-seater coupé.
CONSTRUCTION Carbon-fibre composite.
ENGINE 5,998cc V12.
POWER OUTPUT 651 bhp.
TRANSMISSION Six-speed F1 paddle shift.
SUSPENSION Independent with push-rod shockers.
BRAKES Ceramic composite discs.
MAXIMUM SPEED 363 km/h (226 mph)
0–60 MPH (0–96 KM/H) 3.2 sec
0–100 MPH (0–161 KM/H) 6.6 sec
A.F.C. 4.7 km/l (11 mpg)

FEATHER LIGHT

The Enzo was an automotive lesson in weight saving. The body
panels and tub are composite and carbon-fibre along with the seats,
doors, and even minor switches in the interior. No sound system
was available. Only the McLaren F1 is lighter, but not by much.

LIGHT BRAKES
*Even brakes are
ceramic composite.*

ELECTRIC WING
*Rear spoiler is
computer-controlled.*

FERRARI *458 Italia*

THE 458 IS THE FINEST Ferrari ever. A narcotic cocktail of blistering performance, smooth ride, low-speed tractability, and gorgeous looks won it *Car of the Year* in 2009. There's a long waiting list, used prices are firm, and it's widely considered to be the coolest prancing horse ever made by the Italian company. The super-quick steering takes just two turns lock-to-lock, the dual-clutch, seven-speed automatic gearbox has no delay, and the 4.5-litre V8 is thunderously fast – hitting 161 km/h (100 mph) in just seven seconds. But the 458's ability to trundle round town at low speeds without any temperament makes it different from every other supercar. This is a Ferrari you don't have to suffer to own and one that can genuinely be used every day.

INCREDIBLE ENGINE
The mid-mounted, direct-injection 4.5 V8 develops 125 bhp per litre – a record for a naturally aspirated piston engine. Maximum horsepower is delivered at a screaming 9,000 rpm, and 80 per cent of torque is available at just 3,250 rpm. The Getrag automatic gearbox (shared with the Mercedes SLS) shifts gears in four-tenths of a second. Sixty mph is on the dial in only 3.3 seconds.

RACING INTERIOR
Cabin was designed by Grand Prix champion Michael Schumacher.

COLOURED CALIPERS
Carbon ceramic brakes have five caliper colour choices.

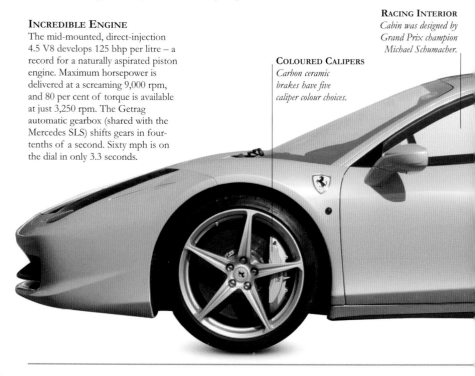

SPECIFICATIONS

MODEL Ferrari 458 Italia (2009)

PRODUCTION N/A

BODY STYLE Mid-engined, two-seater coupé.

CONSTRUCTION Alloy chassis, lightweight panels.

ENGINE 4,499 cc V8.

POWER OUTPUT 562 bhp.

TRANSMISSION Seven-speed, dual-clutch automatic.

SUSPENSION Twin wishbone, double-link.

BRAKES Four-wheel carbon ceramic discs.

MAXIMUM SPEED 325 km/h (202 mph)

0–60 MPH (0–96 KM/H) 3.3 sec

0–100 MPH (0–161 KM/H) 7 sec

A.F.C. 6.4 km/l (15 mpg)

AIR FORCE
Clever vents cool brakes and reduce nose lift.

BEAUTIFUL THING
One of Pininfarina's prettiest designs, the 458 has an alloy chassis bonded using aerospace technology. The underside is flat, and a raft of aerodynamic styling tweaks provide 140 kg (309 lb) of down force at speed. Grip and stability are phenomenal.

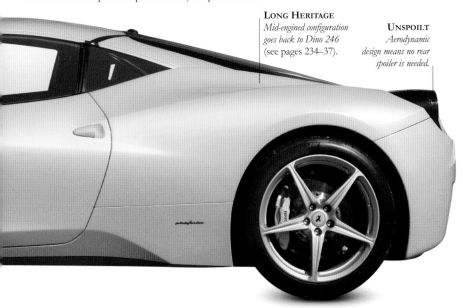

LONG HERITAGE
Mid-engined configuration goes back to Dino 246 (see pages 234–37).

UNSPOILT
Aerodynamic design means no rear spoiler is needed.

FIAT *500D*

WHEN THE FIAT 500 NUOVA appeared in 1957, long-time Fiat designer Dante Giacosa defended it by saying, "However small it might be, an automobile will always be more comfortable than a motor scooter". Today though, the diminutive scoot-about needs no defence, for time has justified Giacosa's faith – over four million 500s and derivatives were produced up to the demise of the Giardiniera estate in 1977. In some senses the Fiat was a mini before the British Mini *(see pages 44–47)*, for the baby Fiat not only appeared two years ahead of its British counterpart, but was also 7.6 cm (3 in) shorter. With its 479cc motor, the original 500 Nuova was rather frantic. 1960 saw it grow to maturity with the launch of the 500D, which was pushed along by its enlarged 499.5cc engine. Now at last the baby Fiat could almost touch 96 km/h (60 mph) without being pushed over the edge of a cliff.

SUNROOF
Some 500s had small fold-back sunroofs. On cabriolets, the fabric roof with plastic rear screen rolled right back.

"SUICIDE" DOORS
You can tell this Fiat is pre-1965 because of the rear-hinged, so-called "suicide" doors. After that the hinges moved to the front in line with more modern practice.

DOORS
The Giardiniera estate kept "suicide" doors until its demise in 1977.

"BONNET"
This houses the petrol tank, battery, and spare wheel, with a little space left for a modest amount of luggage.

HOT FIAT
Carlo Abarth produced a modified and tuned Fiat-Abarth along the lines of the hot Minis created in Britain by John Cooper.

SPECIFICATIONS

MODEL Fiat 500 (1957–77)

PRODUCTION 4 million plus (all models)

BODY STYLES Saloon, cabriolet. Giardiniera estate.

CONSTRUCTION Unitary body/chassis.

ENGINES Two-cylinder air-cooled 479cc or 499.5cc.

POWER OUTPUT 17.5 bhp at 4400 rpm (499.5cc).

TRANSMISSION Four-speed non-synchromesh.

SUSPENSION *Front:* independent, transverse leaf, wishbones; *Rear:* independent semi-trailing arms, coil springs.

BRAKES Hydraulic drums.

MAXIMUM SPEED 95 km/h (59 mph)

0–60 MPH (0–96 KM/H) 32 sec

A.F.C. 19 km/l (53 mpg)

BACK-TO-FRONT
Some rear-engined cars aped front-engined cousins with fake grilles and air intakes. Not the unpretentious Fiat.

CHARMING ITALIAN
This pert little package is big on charm. From any angle the baby Fiat seems to present a happy, smiling disposition. When it comes to parking it is a winner, although accommodation is a bit tight. Two average-sized adults can fit up front, but space in the back is a bit more limited.

INTERIOR

The Fiat 500's interior is minimal but functional. There is no fuel gauge, just a light that illuminates when three-quarters of a gallon remains – enough for another 64 km (40 miles).

REAR SPACE

Realistic back seat permutations are two nippers, one adult sitting sideways, or a large shopping basket.

DRIVING THE 500

The baby Fiat was a fine little driver's car that earned press plaudits for its assured and nimble handling. Although top speed was limited, the car's poise meant you rarely needed to slow down on clear roads.

OPEN-TOP VERSION
Ghia built a Fiat 500-based open beach car called the Jolly. It mimicked pre-war roadsters but is affectionately dubbed the "Noddy" car.

AIR-COOLED REAR
Rear-engined layout, already employed in the Fiat 600 of 1955, saved space by removing the need for a transmission tunnel. The use of an air-cooled engine and only two cylinders in the 500 was a completely new direction for Fiat.

MOTOR
All engines were feisty little devils capable of indefinite flat-out motoring.

FORD *GT40*

TO APPLY THE TERM "SUPERCAR" to the fabled Ford GT40 is to demean it; modern supercars may be uber cool and ferociously fast, but how many of them actually won Le Mans outright? The Ford GT40, though, was not only the ultimate road car but also the ultimate endurance racer of its era, a twin distinction no one else can match. It was so good that arguments are still going on over its nationality. Let us call it a joint design project between the American manufacturer and independent British talent, with a bit of Italian and German input as well. What matters is that it achieved what it was designed for, claiming the classic Le Mans 24-hour race four times in a row. And there is more to the GT40 than its Le Mans legend. You could, if you could afford it, drive around quite legally on public roads in this 322 km/h (200 mph) projectile. Ultimate supercar? No, it is better than that. Ultimate car? Maybe.

CHANGED APPEARANCE

The front section is the easiest way to identify various developments of the GT40. First prototypes had sharp snouts; the squared-off nose, as shown here, first appeared in 1965; the road-going MkIII was smoother, and the end-of-line MkIV rounder and flatter.

SMALL COCKPIT

The cockpit might be cramped, but the GT40's impracticability is all part of its extreme extravagance.

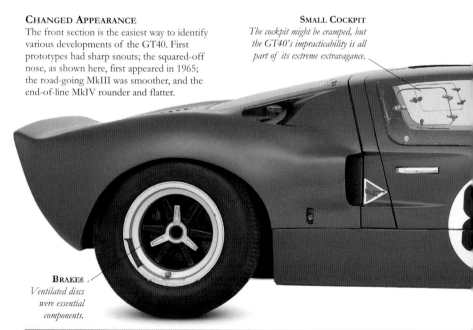

BRAKES
Ventilated discs were essential components.

DOORS
Large doors almost reach centre of roof to ease access.

WINDSCREEN
Panoramic windscreen gave good forward vision.

ENGINE POSITION
Engine slotted almost exactly in middle of car.

DESIGN SECRETS

Design of the GT40 was based on an earlier British Lola. Features such as mid-engined layout with gearbox/transaxle at the rear had by now become standard race-car practice. In Ford's favour were the powerful V8, plenty of bucks, and Henry Ford II's determination to win Le Mans.

ROAD BUMPERS
This is a racer, but road cars had tiny chrome bumpers.

B. BELL

SPECIFICATIONS

MODEL Ford GT40 MkI, II, III, & IV (1964–68)

PRODUCTION 107

BODY STYLE Two-door, two-seat coupé.

CONSTRUCTION Sheet-steel monocoque (honeycomb MkIV), glass-fibre body.

ENGINE Ford V8, 4195cc (MkI), 4727cc (MksI & III), 6997cc (MksII & IV).

POWER OUTPUT From 350 bhp at 7200 rpm (Mk1 4195cc) to 500 bhp at 5000 rpm (MkIV).

TRANSMISSION Transaxle and four- or five-speed ZF gearbox.

SUSPENSION Independent by coil springs and wishbones all round.

BRAKES Ventilated discs all round.

MAXIMUM SPEED 249–322 km/h (155–200 mph, depending on gearing)

0–60 MPH (0–96 KM/H) 4.5 sec

0–100 MPH (0–161 KM/H) 8.5 sec

A.F.C. 4.2–5.7 km/l (12–16 mpg)

WING MIRRORS
Many race cars dispensed with wing mirrors.

EXHAUSTS
Exhaust note rises from gruff bellow to ear-splitting yowl.

TAIL
Lip on tail helped high-speed stability.

VITAL STATISTICS
GT, of course, stands for Grand Touring; 40 for the car's height in inches. Overall length was 4.2 m (13 ft 9 in) width 1.78 m (5 ft 10 in), and unladen weight 832 kg (1,835 lb).

WHEELS
Wheel widths varied depending on racing requirements.

WIND CHEATER
The graceful and muscular shape was penned in Ford's Dearborn design studios. Requirements included a mid-engined layout and aerodynamic efficiency, vital for burning off Ferraris on the straights of Le Mans.

NO-FRILLS CABIN
The GT40's cabin was stark and cramped. Switches and instruments were pure racer, and the low roof line meant that tall drivers literally could not fit in, with the gullwing doors hitting the driver's head.

REAR VISION
Fuzzy slit above engine cover gives just enough rear vision to watch a Ferrari fade away.

STILL WINNING
GT40s can still be seen in retrospective events such as the 1994 Tour de France rally, which the featured car won. The British-owned car proudly displays the British Racing Drivers' Club badge.

VENTS
Ducts helped hot air escape from radiator.

FORD *Thunderbird (1955)*

CHEVY'S 1954 CORVETTE may have been a peach, but anything GM could do, Ford could do better. The '55 T-Bird had none of the 'Vette's glass-fibre nonsense, but a steel body and grunty V8 motor. Plus it was drop-dead gorgeous and offered scores of options, with the luxury of wind-up windows. Nobody was surprised when it outsold the creaky Corvette 24-to-one. But Ford wanted volume and two-seaters weren't everybody's cup of tea, which is why by 1958 the Little Bird became the Big Bird, swollen by four fat armchairs. Nevertheless, as the first of America's top-selling two-seaters, the Thunderbird fired the public's imagination. For the next decade American buyers looking for lively power in a stylish package would greedily devour every Thunderbird going.

NOD TO THE PAST
The styling was very Ford, penned by Bill Boyer and supervised by Frank Hershey. The simple, smooth, and youthful outer wrapping was a huge hit. A rakish long bonnet and short rear deck recalled the 1940s Lincoln Continental.

POWER BULGE
The bonnet needed a bulge to clear the large air cleaners. It was stylish too.

SPECIFICATIONS

MODEL Ford Thunderbird (1955)

PRODUCTION 16,155 (1955)

BODY STYLE Two-door, two-seater convertible.

CONSTRUCTION Steel body and chassis.

ENGINE 292cid V8.

POWER OUTPUT 193 bhp.

TRANSMISSION Three-speed manual with optional overdrive, optional three-speed Ford-O-Matic automatic.

SUSPENSION *Front:* independent coil springs; *Rear:* leaf springs with live axle.

BRAKES Front and rear drums.

MAXIMUM SPEED 169–201 km/h (105–125 mph)

0–60 MPH (0–96 KM/H) 7–11 sec

A.F.C. 6 km/l (17 mpg)

INTERIOR

Luxury options made the Thunderbird an easy-going companion. On the list were power steering, windows, and brakes, automatic transmission, and even electric seats and a power-assisted top. At $100, the push-button radio was more expensive than power steering.

COCKPIT

With the top up, heat from the transmission made for a hot cockpit; ventilation flaps were introduced on '56 and '57 models.

SMOOTH LINES

For 1955, this was an uncharacteristically clean design and attracted 16,155 buyers in its first year of production.

LENGTH

Hardly short, the Little Bird measured 4.4 m (175 in).

CLEARANCE

Road clearance was limited at just 12.7 cm (5 in).

ENGINE
The T-Bird's motor was the new cast-iron OHV 292cid V8 with dual exhausts and four-barrel Holley carb. Compared to the 'Vette's ancient six, the T-Bird's mill offered serious shove. Depending on the state of tune, a very hot T-Bird could hit 60 in seven seconds.

WINDSCREEN
The aeronautical windscreen profile is beautifully simple.

SUCCESSFUL BLOCK
The Thunderbird's V8 played a major role in the car's success.

OVERHEAD VIEW
This overhead shot shows that the T-Bird had a bright and spirited personality. Today, the T-Bird is a fiercely prized symbol of American Fifties utopia. The '55–'57 Thunderbirds are the most coveted – the model turned into a four-seater in 1958.

SIMPLE STYLING
Apart from the rather too prominent exhausts, the rear end is remarkably uncluttered. Hardtops were standard fare but soft-tops could be ordered as a factory option.

POWER STEERING
*Power steering would only
cost the buyer a bargain $98.*

ENGINE OUTPUT
*Power output ranged
from 212 to 300 horses.
Buyers could beautify
their motors with a $25
chrome dress-up kit.*

T-BIRD NAME
The Thunderbird name was chosen
after a south-west Native American
god who brought rain and prosperity.
Star owners included the movie
actresses Debbie Reynolds, Marilyn
Monroe, and Jayne Mansfield.

FORD *Fairlane 500 Skyliner*

FORD RAISED THE ROOF IN '57 with their glitziest range ever, and the "Retrac" was a party piece. The world's only mass-produced retractable hardtop debuted at the New York Show of '56 and the first production version was presented to a bemused President Eisenhower in '57. The Skyliner's balletic routine was the most talked-about gadget for years and filled Ford showrooms with thousands of gawping customers. Surprisingly reliable and actuated by a single switch, the Retrac's roof had 185 m (610 ft) of wiring, three drive motors, and a feast of electrical hardware. But showmanship apart, the Skyliner was pricey and had precious little boot space or leg room. By '59 the novelty had worn off and division chief Robert McNamara's desire to end expensive "gimmick engineering" led to the wackiest car ever to come out of Dearborn being axed in 1960.

DECLINING NUMBERS

The Skyliner lived for three years but was never a volume seller. Buyers may have thought it neat, but they were justifiably anxious about the roof mechanism's reliability. Just under 21,000 were sold in '57, less than 15,000 in '58, and a miserly 12,915 found buyers in '59.

ENGINE CHOICE

The Skyliner could be specified with four different V8s ranging from 272 to 352cid.

GLASS EXTRA

Options included tinted glass, power windows, power seat, and Styleton two-tone paint.

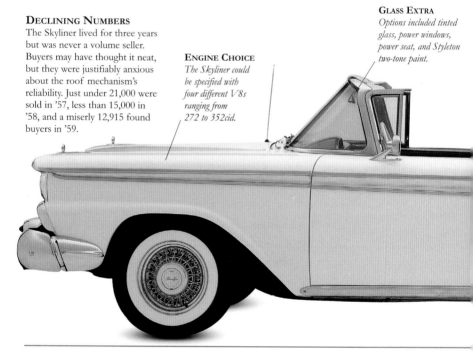

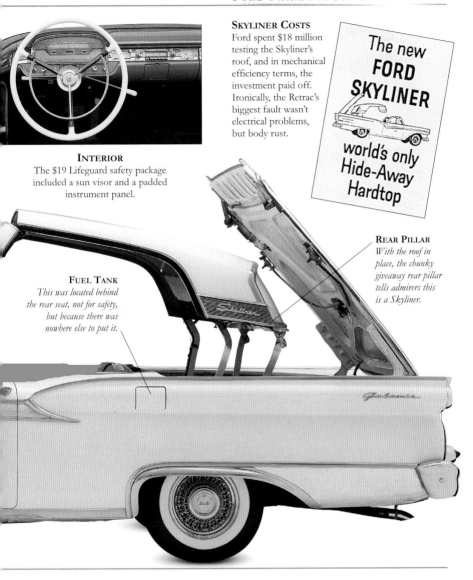

SKYLINER COSTS
Ford spent $18 million testing the Skyliner's roof, and in mechanical efficiency terms, the investment paid off. Ironically, the Retrac's biggest fault wasn't electrical problems, but body rust.

The new
**FORD
SKYLINER**
world's only
**Hide-Away
Hardtop**

INTERIOR
The $19 Lifeguard safety package included a sun visor and a padded instrument panel.

REAR PILLAR
With the roof in place, the chunky giveaway rear pillar tells admirers this is a Skyliner.

FUEL TANK
This was located behind the rear seat, not for safety, but because there was nowhere else to put it.

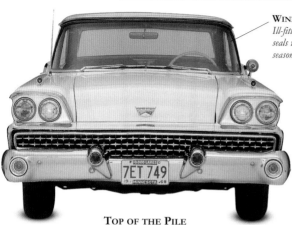

WINDSCREEN
Ill-fitting window seals were an all-season annoyance.

CHASSIS
Chassis had to be modified to leave room for the top's control linkage.

TOP OF THE PILE
At two tonnes and $3,138, the Skyliner was the heaviest, priciest, and least practical Ford in the range. The Skyliner's standard power was a 292cid V8, but this model contains the top-spec Thunderbird 352cid Special V8 with 300 bhp.

SUSPENSION
Though a particularly heavy car, rear suspension was by standard leaf springs.

BOOT LID
Boot lid hinged from the rear and folded down over the retracted roof.

STYLING
The boot sat higher on Skyliners. Large circular rear lights were very Thunderbird and became a modern Ford trademark.

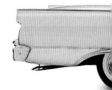

REAR VIEW
Fins were down for '59, but missile-shaped pressings on the higher rear wings were a neat touch to hide all that moving metalwork. Supposedly a mid-sized car, the Fairlane was the first of the long, low Fords.

HOOD UP

With the roof up, the optional Polar-Aire air-conditioning made sense. Other extras that could be specified included tinted glass and, most important for the Retrac, a 70-amp heavy-duty battery. Skyliners came with a comprehensive troubleshooting instruction booklet along with a very slow and ponderous manual back-up system.

SPECIFICATIONS

MODEL Ford Fairlane 500 Galaxie Skyliner Retractable (1959)

PRODUCTION 12,915 (1959)

BODY STYLE Two-door hardtop with retractable roof.

CONSTRUCTION Steel body and chassis.

ENGINES 272cid, 292cid, 312cid, 352cid V8s.

POWER OUTPUT 190–300 bhp.

TRANSMISSION Three-speed manual, optional three-speed Cruise-O-Matic automatic.

SUSPENSION *Front:* coil springs; *Rear:* leaf springs.

BRAKES Front and rear drums.

MAXIMUM SPEED 169 km/h (105 mph)

0–60 MPH (0–96 KM/H) 10.6 sec

A.F.C. 5.4 km/l (15.3 mpg)

HOOD PROCEDURE

A switch on the steering column started three motors that opened the rear deck. Another motor unlocked the top, while a further motor hoisted the roof and sent it back to the open boot space. A separate servo then lowered the rear deck back into place. It all took just one minute, but had to be done with the engine running.

FORD *Galaxie 500XL Sunliner*

IN '62, FORD WERE SELLING their range as "America's liveliest, most carefree cars". And leading the lively look was the bright-as-a-button new Galaxie. This was General Manager Lee Iacocca's third year at the helm and he was pitching for the young-guy market with speed and muscle. Clean cut, sleek, and low, the Galaxie range was just what the boys wanted and it drove Ford into a new era. The new-for-'62 500XL was a real piece, with bucket seats, floor shift, a machine-turned instrument panel, and the option of a brutish 406cid V8. XL stood for "extra lively", making the 500 one of the first cars to kick off Ford's new Total Performance sales campaign. The 500XL Sunliner Convertible was billed as a sporting rag-top and cost an eminently reasonable $3,350. Engines were mighty, rising from 292 through 390 to 406cid V8s, with a Borg-Warner stick-shift four-speed option. Ford learnt an important lesson from this car. Those big, in-yer-face engines clothed in large, luxurious bodies would become seriously hip.

BELTS
Front seat belts were an option.

TREND-SETTER
The slab-sided Galaxie body was completely new for '62 and would set something of a styling trend for larger cars. Lines may have been flat and unadorned, but buyers could choose from 13 colours and 21 jaunty two-tones.

PADDED DASH
The dashboard was padded and colour-matched the exterior.

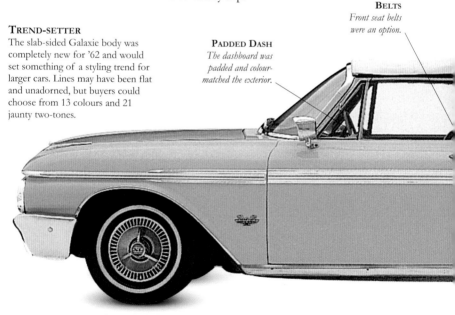

SALES BROCHURE

"This year, more than ever before, Galaxie styling is the envy of the industry." Subjective sales literature maybe, but Ford's restyled Galaxies were a real success, and the new XL series offered peak performance plus the top trim level of the 500.

SPECIFICATIONS

MODEL Ford Galaxie 500XL Sunliner Convertible (1962)

PRODUCTION 13,183 (1962)

BODY STYLE Two-door convertible.

CONSTRUCTION Steel body and chassis.

ENGINES 292cid, 352cid, 390cid, 406cid V8s.

POWER OUTPUT 170–405 bhp.

TRANSMISSION Three-speed Cruise-O-Matic automatic, optional four-speed manual.

SUSPENSION *Front:* coil springs; *Rear:* leaf springs.

BRAKES Front and rear drums.

MAXIMUM SPEED 174–225 km/h (108–140 mph)

0–60 MPH (0–96 KM/H) 7.6–14.2 sec

A.F.C. 5.7–6.4 km/l (16–18 mpg)

ROOF
Glass-fibre "blankets" insulated the roof.

TOP UP
Unlike this example, the rarest Sunliners had a wind-cheating Starlift hardtop, which was not on the options list.

STYLISH CHROME
The arrow-straight side flash is a far cry from the florid sweepspears that adorned most Fifties models.

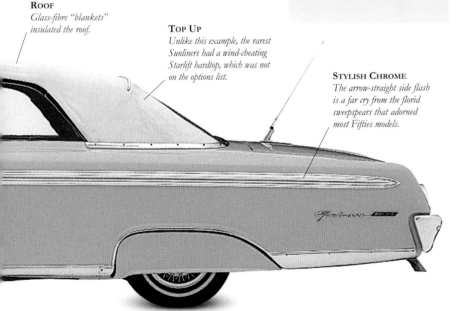

INTERIOR

The interior was plush and palatial, with Mylar-trimmed, deep-pleated buckets flanking the centre console. Seats could be adjusted four ways manually and six ways electronically.

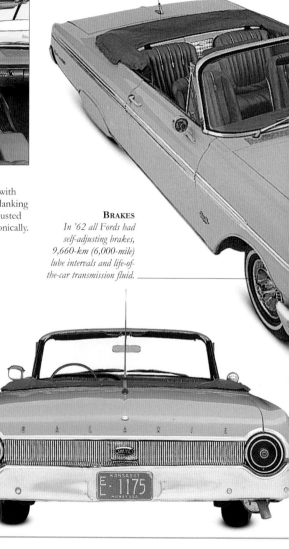

BRAKES

In '62 all Fords had self-adjusting brakes, 9,660-km (6,000-mile) lube intervals and life-of-the-car transmission fluid.

LIGHTS

Large, round, rear-light cluster aped the T-Bird and appeared on the Falcon as well as the Fairlane, also debuting in 1962.

REAR ASPECT

Fuel filler-cap lurks behind the central hinged section of the anodized beauty panel. The panel itself highlights the car's width. The hardtop version of the 500XL Sunliner was the Club Victoria, $250 cheaper than the convertible and twice as popular, with 28,000 manufactured in '62.

MIRROR-LIGHT
The spotlight-mirror was a factory option; on a clear day, the light could emit a beam 800 metres (½ mile) ahead.

BODY INSULATION
The Galaxie had an especially quiet ride because it was soundproofed at various points. Sound-absorbent mastic was applied to the inside surfaces of the doors, bonnet, boot lid, wings, and quarter panels.

GALAXIE PERFORMANCE
The Galaxies of '62 marked Ford boss Lee Iacocca's first sortie into the performance-obsessed youth market, which two years later would blossom into the legendary Mustang *(see pages 278–85)*. It was an inspired marketing gamble that took Ford products through the Sixties with huge success in both showrooms and on the race-track.

ENGINE
Stock Galaxies lumbered around with a 223cid six or 292cid V8. The 500XL could choose from a range of Thunderbird V8s that included the 390cid Special, as here, and a 405 bhp 406cid V8 with triple Holley carbs, which could be ordered for $379.

CHASSIS
Chassis was made up of wide-contoured frame with double-channel side rails.

FORD *Thunderbird* *(1962)*

IT WAS NO ACCIDENT THAT THE third-generation T-Bird looked like it was fired from a rocket silo. Designer Bill Boyer wanted the new prodigy to have "an aircraft and missile-like shape", a subtext that wasn't lost on an American public vexed by the Cuban crisis and Khrushchev's declaration of an increase in Soviet military spending. The Sports Roadster model was the finest incarnation of the '61–'63 Thunderbird. With Kelsey-Hayes wire wheels and a two-seater glass-fibre tonneau, it was one of the most glamorous cars on the block and one of the most exclusive. Virile, vast, and expensive, the Big Bird showed that Detroit still wasn't disposed to making smaller, cheaper cars. GM even impudently asserted that "a good used car is the only answer to America's need for cheap transportation". And anyway, building cars that looked and went like ballistic missiles was far more interesting and profitable.

PRETTY CONVERTIBLE
With the bonnet down, the Big Bird was one of the most attractive and stiffest convertibles Ford had ever made. The heavy unitary-construction body allowed precious few shakes, rattles, and rolls. *Motor Trend* magazine said: "Ford's plush style-setter has plenty of faults... but it's still the classic example of the prestige car."

TILT WHEEL
T-Bird drivers weren't that young, and a Swing-Away steering wheel aided access for the more corpulent driver.

WHEELS
Lesser T-Birds could opt for the Roadster's wire wheels at $343.

CHASSIS

Construction was "dual-unitized", with separate front and rear sections welded together at the cowl.

ROOF FUN

With the top down, the streamlined tonneau made the Sports Roadster sleek enough to echo the '55 two-seater Thunderbird *(see pages 262–65)*.

SPECIFICATIONS

MODEL Ford Thunderbird Sports Roadster (1962)

PRODUCTION 455 (1962)

BODY STYLE Two-door, two/four-seater convertible.

CONSTRUCTION Steel body and chassis.

ENGINE 390cid V8.

POWER OUTPUT 330–340 bhp.

TRANSMISSION Three-speed Cruise-O-Matic automatic.

SUSPENSION *Front:* upper and lower A-arms and coil springs; *Rear:* leaf springs with live axle.

BRAKES Front and rear drums.

MAXIMUM SPEED 187–201 km/h (116–125 mph)

0–60 MPH (0–96 KM/H) 9.7–12.4 sec

A.F.C. 3.9–7.1 km/l (11–20 mpg)

DECORATION

Three sets of five cast-chrome slash marks unmistakably suggest total power.

BODY CREASE

Odd styling crease ran from wing to door and is the model's least becoming feature.

OVERHANG

Rear overhang was prodigious, but parking could be mastered by using the rear fin as a marker.

INTERIOR
Aircraft imagery in the controls is obvious. The interior was designed around a prominent centre console that split the cabin into two separate cockpits, delineating positions of driver and passenger.

ADDED EXTRAS
Tinted glass, power seats and windows, and AM/FM radio were the most popular options.

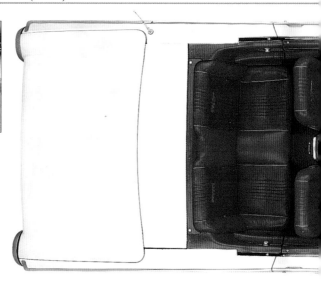

OVERHEAD VIEW
The Sports Roadster could also be a full four-seater. Trouble was there was no space in the boot for the tonneau, so it had to stay at home. The large tonneau panel came off easily but required two people to handle it.

FRONT ASPECT
The front bears an uncanny resemblance to the British Ford Corsair, which is neither surprising nor coincidental, since the Corsair was also made by Uncle Henry. This third-generation T-Bird was warmly received and sold well.

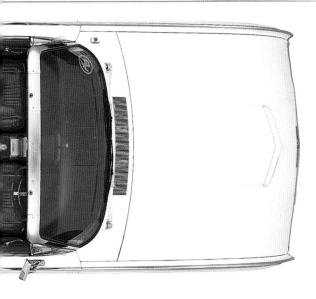

DIVINE DESIGN
Sales literature suggested
that the T-Bird was the result
of the combined efforts
of Ford and God.

CABIN DESIGN
*Interior designer Art Querfield
spent more time on the T-
Bird's cabin than on any other
car in his 40 years at Ford.*

CLEANER REAR
Ford cleaned up the rear of their
prestige offering after the demise
of the '58 to '60 Squarebird.
Lights were a simple
rounded cluster and the
bumper was straight
and wide.

COLOURS
*18 single shades
or 24 two-tone
combinations
were offered.*

FORD *Mustang (1965)*

THIS ONE HIT THE GROUND RUNNING – galloping in fact, for the Mustang rewrote the sales record books soon after it burst on to the market in April 1964. It really broke the mould, for it was from the Mustang that the term "pony car" was derived to describe a new breed of sporty "compacts". The concept of an inexpensive sports car for the masses is credited to dynamic young Ford vice-president, Lee Iacocca. In realization, the Mustang was more than classless, almost universal in appeal. Its extensive options list meant there was a flavour to suit every taste. There was a Mustang for mums, sons, daughters, husbands, even young-at-heart grandparents. Celebrities who could afford a ranch full of thoroughbred race horses and a garage full of Italian exotics were also proud to tool around in Mustangs. Why, this car's a democrat.

MASS APPEAL
The Mustang burst into the history books almost the moment it was unveiled to the public in Spring 1964. At one stroke it revived the freedom of spirit of the early sporting Thunderbirds and brought sports car motoring to the masses.

AERIAL
Push-button radio and antenna were all part of the options list.

DOORS
Long doors helped entry and exit for rear passengers.

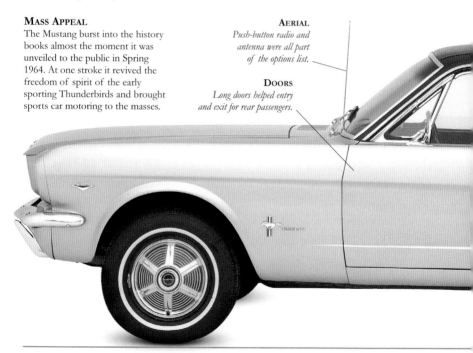

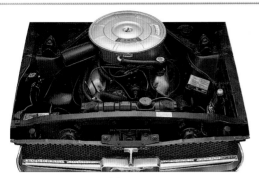

ENGINE

Mustangs were offered with the option of V8 (289cid pictured) or six-cylinder engines; eights outsold sixes two-to-one in 1964–68. Customers could thus buy the car with 100 bhp or have 400 bhp sports car performance.

SPECIFICATIONS

MODEL Ford Mustang (1964–68)

PRODUCTION 2,077,826

BODY STYLES Two-door, four-seat hardtop, fastback, convertible.

CONSTRUCTION Unitary chassis/body.

ENGINES Six-cylinder 170cid to 428cid V8. Featured car: 289cid V8.

POWER OUTPUT 195–250 bhp at 4000–4800 rpm or 271 bhp at 6000 rpm (289cid).

TRANSMISSION Three- or four-speed manual or three-speed automatic.

SUSPENSION Independent front by coil springs and wishbones; semi-elliptic leaf springs at rear.

BRAKES Drums; discs optional at front.

MAXIMUM SPEED 177–204 km/h (110–127 mph) (289cid)

0–60 MPH (0–96 KM/H) 6.1 sec (289cid)

0–100 MPH (0–161 KM/H) 19.7 sec

A.F.C. 4.6 km/l (13 mpg)

PILLARLESS COUPÉ
Both front and rear side windows wound completely out of sight.

WHEEL OPTIONS
Myriad options included smaller wheels, wider tyres, wire wheel-covers, and knock-off style hub embellishers.

INTERIOR
The first Mustangs shared their
instrument layout with more mundane
Ford Falcons, but in a padded dash.
The plastic interior is a little tacky, but
at the price no one was going to
complain. The sports wheel was
a standard 1965 fitment.

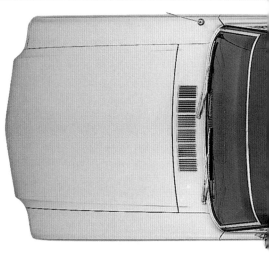

WINDSCREEN
*Banded, tinted
windscreen was
another option.*

PROTOTYPE ORIGINS
The Mustang I prototype
of 1962 was a V4 mid-
engined two-seater – pretty
but too exotic. The four-
seater Mustang II show car
debuted at the US Grand
Prix in 1963, and its
success paved the way for
the production Mustang,
which to this day is still the
fastest selling Ford ever.

BRAKES
*Front discs were
a new option
for 1965.*

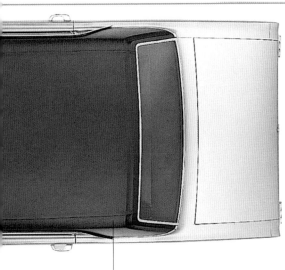

OVERHEAD VIEW

This bird's-eye view of the Mustang shows the sense of its proportions, with a box for the engine, the people, and their luggage. Interior space was maximized by doing away with Detroit's bulging, and often florid, outer panels. The Mustang's almost understated styling was a breath of fresh air.

V-SIGN

The 289 cubic inch, cast-iron V8 engine was a glamorous power unit, seeing service in the iconic AC Cobra, Sunbeam Tiger, and TVR Griffiths.

ROOF
Popular vinyl-covered roof option on the hardtop simulates the convertible.

CONSUMER CHOICE

The Mustang could be as cheap or expensive as you liked. "The Mustang is designed to be designed by you" gushed an early sales brochure. From an entry price of $2,368, you could simply tick the option boxes to turn your "personal" car into a hot-rod costing more than double that.

SUSPENSION
Harder suspension and handling kits could be ordered as an option.

FORD *Shelby Mustang GT500 (1967)*

LOOKING BACK FROM OUR ERA of weedy political correctness, it's amazing to remember a time when you could buy this sort of stomach-churning horsepower straight from the showroom floor. What's more, if you couldn't afford to buy it, you could borrow it for the weekend from your local Hertz rent-a-car. The fact is that the American public loved the grunt, the image, and the Carroll Shelby Cobra connection. Ford's advertising slogan went straight to the point – Shelby Mustangs were "*The* Road Cars". With 289 and 428cid V8s, they were blisteringly quick and kings of both street and strip. By '67 they were civilized too, with options like factory air and power steering, as well as lots of gauges, a wood-rim Shelby wheel, and that all-important 140 mph (225 km/h) speedo. The little Pony Mustang had grown into a thundering stallion.

STEERING WHEEL
The wood-rim steering wheel came with the Shelby package.

THE SHELBY IN '67
'67 Shelbys had a larger bonnet scoop than previous models, plus a custom-built glass-fibre front to complement the stock Mustang's new longer bonnet. Shelbys were a big hit in '67, with 1,175 350s and 2,048 500s sold. Prices were also about 15 per cent cheaper than in '66.

500 NAME
GT500 name was arbitrary and did not refer to power.

LOCK PINS
Racing-style lock pins were standard on the bonnet.

SHELBY PLATE

Shelby gave the early Mustangs his special treatment in a factory in Los Angeles. Later cars were built in Michigan. Shelby delivered the first batch to Hertz the day before a huge ice storm. The brakes proved too sharp and 20 cars were written off.

SPECIFICATIONS

MODEL Ford Shelby Cobra Mustang GT500 (1967)

PRODUCTION 2,048 (1967)

BODY STYLE Two-door, four-seater coupé.

CONSTRUCTION Steel unitary body.

ENGINE 428cid V8.

POWER OUTPUT 360 bhp.

TRANSMISSION Four-speed manual, three-speed automatic.

SUSPENSION *Front:* coil springs; *Rear:* leaf springs.

BRAKES Front discs, rear drums.

MAXIMUM SPEED 212 km/h (132 mph)

0–60 MPH (0–96 KM/H) 6.8 sec

A.F.C. 4.6 km/l (13 mpg)

HERTZ FUN
There are tales of rented Shelbys being returned with bald tyres and evidence of racing numbers on the doors.

COBRA REBIRTH
At the end of '67, cars were renamed Shelby Cobras, but Ford still handled all promotion and advertising.

SIDE SCOOPS
Scoops acted as interior air extractors for the Shelby.

REAR DECK
Rear deck was made of glass-fibre to save weight.

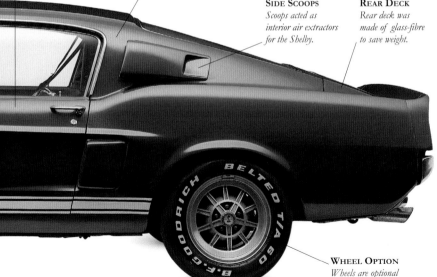

WHEEL OPTION
Wheels are optional Kelsey-Hayes Magstars.

PRACTICAL SEATING

All GT350s and 500s boasted the standard and very practical Mustang fold-down rear seat along with Shelby's own padded roll-bar. Shelbys came in fastback only; there were no notchbacks and convertibles were only available from '68.

SUSPENSION
Shelby's springing was similar to the Mustang with front sway bar, stiff springs, and Gabriel shocks.

FUEL CONSUMPTION
Thirsty 428cid V8 meant that only 4.6 km/l (13 mpg) was possible.

COBRA ORIGINS
428cid V8 started life in the original AC Cobra.

THE 500'S BLOCK

The GT500 came with the 428 Police Interceptor unit with two Holley four-barrel carbs. Oval, finned aluminium open-element air cleaner and cast-aluminium valve covers were unique to the big-block Shelby.

CENTRE LIGHTS
The standard centre-gri high-beam headlights were forced to the sides some states because of federal legislation.

TACHOMETER
The standard tachometer red-lined at 8000 rpm.

INTERIOR
Stewart-Warner oil and amp gauges and a tachometer were standard fittings. Two interior colours were available – parchment and black. Interior decor was brushed aluminium with moulded door panels and courtesy lamps.

BRAKES
The rear drum brakes were assisted at the front by more efficient discs.

POWER REFINEMENTS
The introduction of power-assisted steering and brakes in the '67 model meant that the once rough-riding Shelby had transformed into a luxury slingshot that would soon become an icon.

LIGHTS
For the Shelby, the Mustang's rear lights were replaced with the '65 T-Bird's sequential lights.

GORDON KEEBLE *GT*

IN 1960, THIS WAS THE MOST ELECTRIFYING CAR the British magazine *Autocar & Motor* had ever tested. Designed by Giugiaro in Italy and built in an aircraft hanger in Southampton, it boasted good looks, a glass-fibre body, and a 5.4-litre, 300 bhp V8 Chevrolet Corvette engine. But, despite plenty of publicity, good looks, epic performance, and a glamorous clientele, the Gordon Keeble was a commercial disaster, with only 104 built. "The car built to aircraft standards", read the advertising copy. And time has proved the Keeble's integrity; a space-frame chassis, rust-proof body, and that unburstable V8 has meant that over 90 Gordons have survived, with 60 still regularly used today. Born in an era where beauty mattered more than balance sheets, the Gordon Keeble failed for two reasons. Firstly the workers could not make enough of them, and secondly the management forgot to put a profit margin in the price. How the motor industry has changed...

STYLE
For a '60s' design, the Gordon Keeble is crisp, clean, and timeless.

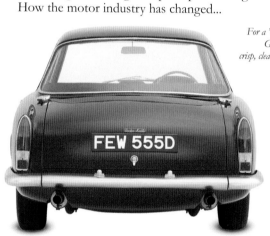

SPACE FRAME
The prototype space-frame chassis was a composite skeleton of square tubes. It was flown to France, then overland to Turin, where Giugiaro fitted a handsome grp body.

BARGAIN CLASSIC
Like most classic cars, the Gordon Keeble has fallen in price since the late-'80s. In the UK, good examples can be bought for £10,000, or half their 1988 value.

BUMPERS
The Keeble's delicate three-piece chrome bumpers were specially hand-made.

WINDOWS
Electric windows used the same motors as the Rolls-Royce Silver Shadow.

ENGINE
The small block Sting Ray engine delivered a massive 300 bhp of high-compression power.

YOUNG DESIGNER
Only 21 when he designed the car, Giugiaro gave the bonnet a dummy intake scoop and fashionably raked twin headlights. The roof was lengthened and the slant of the C-pillar decreased to give wider glass areas and maximum visibility.

HIGH-QUALITY BODY
In its day the Keeble's hand-finished, glass-reinforced plastic body was among the best.

SPECIFICATIONS

MODEL Gordon Keeble GT (1964–67)

PRODUCTION 104

BODY STYLE Four-seater glass-fibre GT.

CONSTRUCTION Multi-tubular chassis frame, grp body.

ENGINE 5.4-litre V8.

POWER OUTPUT 300 bhp at 5000 rpm.

TRANSMISSION Four-speed all-synchro.

SUSPENSION Independent front, De Dion rear end.

BRAKES Four-wheel disc.

MAXIMUM SPEED 227 km/h (141 mph)

0–60 MPH (0–96 KM/H) 7.5 sec

0–100 MPH (0–161 KM/H) 13.3 sec

A.F.C. 5 km/l (14 mpg)

HOLDEN *FX*

AT THE END OF WORLD WAR II, Australia had a problem – an acute shortage of cars and a newly civilized army with money to burn. Loaded with Government handouts, General Motors-Holden came up with a four-door, six-cylinder, six-seater that would become an Australian legend on wheels. Launched in 1948, the 48-215, more generally known as the FX, was Australia's Morris Minor *(see pages 378–81)*. Tubby, conventional, and as big as a Buick, it had a sweet, torquey engine, steel monocoque body, hydraulic brakes, and a three-speed column shift. Light and functional, the FX so impressed Lord Nuffield (of Morris fame) with its uncomplicated efficiency that he had one shipped to England for his engineers to pull apart. The Australians did not care about the FX's humble underpinnings and bought 120,000 with grateful enthusiasm.

CLEARANCE
High ground clearance was especially designed for bad roads.

US INFLUENCE

The "Humpy Holden" was a warmed-over pre-war design for a small Chevrolet sedan that General Motors US had created in 1938. A Detroit-Adelaide collaboration, the FX eventually emerged as a plain shape that would not date. Australians still speak of the FX in hallowed tones, remembering it as one of the decade's most reliable cars.

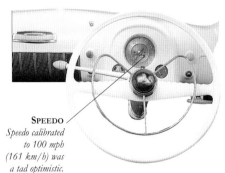

SPEEDO

Speedo calibrated to 100 mph (161 km/h) was a tad optimistic.

DASHBOARD

The dash echoes the Australian culture for utilitarianism, with centre speedo, two occasional gauges, three-speed column change, and only five ancillary switches. The umbrella handbrake and chrome horn-ring were hangovers from Detroit design influences.

BODY FLEX

Taxi drivers complained of body flexing – doors could spring open on corners.

BODY SHELL

Body was dust-proof, which helped in the hot Australian climate.

ECONOMY

Post-war fuel shortages meant that the Holden was parsimonious.

LIGHTS
Simple and unadorned, the FX had no indicators or sidelights, just a six-volt electrical system with a single rear lamp.

REAR WING STYLING
The Holden's rear wing line was cut into the rear doors but was much milder than Detroit's styling men would have liked. Rear wing spats were fitted to make the car look lower and sleeker. Endlessly practical, the FX had a cavernous luggage compartment.

ENGINE
Power came from a sturdy 2170cc cast-iron straight-six, with an integral block and crankcase, push-rod overhead valves, and a single-barrel downdraught Stromberg carburettor.

FRONT ASPECT
Recumbent lion bonnet
mascot lent the FX an
illusion of pedigree. In
reality, Holden had no
bloodline at all, but that
didn't matter as it went on
to become the standard
transport of the Australian
middle classes.

OUTPUT
*The engine
developed a
modest 60 bhp.*

SPECIFICATIONS

MODEL Holden 48-215 FX (1948–53)

PRODUCTION 120,402

BODY STYLE Six-seater, four-door
family saloon.

CONSTRUCTION All-steel Aerobilt
monocoque body.

ENGINE Six-cylinder cast-iron 2170cc.

POWER OUTPUT 60 bhp at 4500 rpm.

TRANSMISSION Three-speed manual.

SUSPENSION *Front:* coil and wishbone;
Rear: leaf spring live axle.

BRAKES Four-wheel hydraulic drums.

MAXIMUM SPEED 117 km/h (73 mph)

0–60 MPH (0–96 KM/H) 27.7 sec

A.F.C. 11 km/l (30 mpg)

SUSPENSION
*The Holden was too powerful
for its suspension and many
ended up on their roofs.*

BROCHURES
General Motors-Holden started
life as a saddlery and leather goods
manufacturer, later diversifying
into car body builders.

HUDSON *Super Six*

IN 1948, HUDSON'S FUTURE could not have looked brighter. The feisty independent was one of the first with an all-new post-war design. Under the guidance of Frank Spring, the new Hudson Super Six not only looked stunning, it bristled with innovation. The key was its revolutionary "step-down" design, based on a unitary construction, with the floor pan suspended from the bottom of the chassis frame. The Hudson was lower than its rivals, handled with ground-hugging confidence, and with its gutsy six-cylinder engine, outpaced virtually all competitors. In 1951, it evolved into the Hudson Hornet, dominating American stock car racing from 1951 to 1954. But the complex design could not adapt to the rampant demand for yearly revision; the 1953 car looked much like the 1948, and in 1954 Hudson merged with Nash, disappearing for good in 1957.

AERODYNAMIC PROFILE
It is the smooth beauty of the profile that really marks the Hudson out. The design team was led by Frank Spring, a long-time Hudson fixture, whose unusual blend of talents combined styling and engineering. His experience in aeroplane design is evident in the Hudson's aerodynamics.

HEIGHT
Only 1.53 m (60.4 in) high, the Super Six was lower than its contemporaries.

LOW RIDER
Chassis frame ran outside the rear wheels, serving as "invisible side bumpers".

UNDER THE BONNET

The gutsy new 262cid six arrived in 1948 and made the Hudson one of the swiftest cars on America's roads.

SPECIFICATIONS

MODEL Hudson Super Six (1948–51)

PRODUCTION 180,499

BODY STYLES Four-door sedan, Brougham two-door sedan, Club coupé, hardtop coupé, two-door Brougham convertible.

CONSTRUCTION Unitary chassis/body.

ENGINE 262cid L-head straight-six.

POWER OUTPUT 121 bhp at 4000 rpm.

TRANSMISSION Three-speed manual, optional overdrive; semi-automatic.

SUSPENSION *Front:* independent, wishbones, coil springs, telescopic dampers, anti-roll bar; *Rear:* live-axle, semi-elliptic leaf springs, telescopic dampers, anti-roll bar.

BRAKES Hydraulic drums all round.

MAXIMUM SPEED 145 km/h (90 mph)

0–60 MPH (0–96 KM/H) 14–18 sec (depending on transmission)

A.F.C. 4.2–6.4 km/l (12–18 mpg)

SPLIT SCREEN

Each segment of the split screen was well curved for semi-wrap-around effect and good visibility.

POST-WAR PIONEER

Along with Studebaker, the 1948 Hudson was one of the very first all-new post-war designs.

SUSPENSION

Front suspension was by wishbones, coil springs, and telescopic dampers.

HUDSON *Hornet*

HUDSON DID THEIR BEST IN '54 to clean up their aged 1948 body. Smoother flanks and a lower, wider frontal aspect helped, along with a new dash and brighter fabrics and vinyls. And at long last the windscreen was one-piece. Mechanically it wasn't bad either. In fact, some say the last Step-Down was the best ever. With the straight-six came a Twin-H power option, a hot camshaft, and an alloy head that could crank out 170 bhp; it was promptly dubbed "The Fabulous Hornet". The problem was that everybody had V8s, and by mid-'54 Hudson had haemorrhaged over $6 million. In April that year, Hudson, who'd been around since 1909, were swallowed up by the Nash-Kelvinator Corporation. Yet the Hornet has been rightly recognized as a milestone car and one of the quickest sixes of the era. If Hudson are to be remembered for anything, it should be for their innovative engineers, who could wring the best from ancient designs and tiny budgets.

POWERFUL STEP-DOWNS
These Hudsons were known as Step-Downs because you literally stepped down into the car. Among the fastest cars of the Fifties, they boasted above-average power and crisp handling.

ENGINE
Amazingly, Hudson never offered V8 power, which was to hasten their downfall.

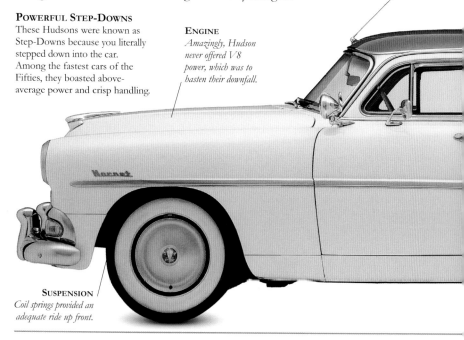

SUSPENSION
Coil springs provided an adequate ride up front.

SPECIFICATIONS

MODEL Hudson Hornet 7D (1954)

PRODUCTION 24,833 (1954 Hornets)

BODY STYLES Two-door coupé or convertible, four-door sedan.

CONSTRUCTION Steel body and chassis.

ENGINE 308cid straight-six.

POWER OUTPUT 160–170 bhp.

TRANSMISSION Three-speed manual, optional Hydra-Matic automatic.

SUSPENSION *Front:* coil springs; *Rear:* leaf springs.

BRAKES Front and rear drums.

MAXIMUM SPEED 177 km/h (110 mph)

0–60 MPH (0–96 KM/H) 12 sec

A.F.C. 6 km/l (17 mpg)

RACING HORNETS

NASCAR devotees watched many a Hudson trounce the competition, winning 22 out of 37 major races in '53 alone. Advertising copy made much of Hudson's racing success and the Hornet was "powered to outperform them all!".

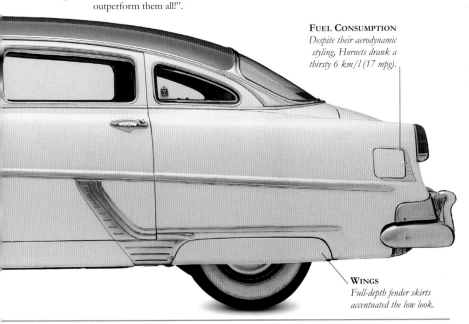

FUEL CONSUMPTION
Despite their aerodynamic styling, Hornets drank a thirsty 6 km/l (17 mpg).

WINGS
Full-depth fender skirts accentuated the low look.

JAGUAR *XK120*

A CAR-STARVED BRITAIN, still trundling around in perpendicular, pre-war hangover motors, glimpsed the future in October of 1948 at the Earl's Court Motor Show in London. The star of the show was the Jaguar Super Sports. It was sensational to look at from any angle, with a purity of line that did not need chrome embellishment. It was also sensationally fast; in production as the Jaguar XK120 it would soon be proven that 120 really did stand for 120 mph (193 km/h), making it the fastest standard production car in the world. The only trouble was that you could not actually buy one. The XK120 was originally planned as a short production-run, prestige show-stopper, but overwhelming interest at the 1948 show changed all that. Hand-built alloy-bodied cars dribbled out of the Jaguar factory in 1949 and you needed a name like Clark Gable to get your hands on one. Tooling was ready in 1950 and production really took off. Today the XK120 is a platinum-plated investment.

FIXED-HEAD HEAVEN
Many rate the fixed-head coupé as the most gorgeous of all XK120s, with a roof line and teardrop window reminiscent of the beautiful Bugatti Type 57SC Atlantic. The fixed-head model did not appear until March 1951 and is much rarer than the roadster.

COCKPIT
The cockpit was a little cosy – if not downright cramped.

TYRES
Skinny cross-ply tyres gave more thrills than needed on hard cornering.

EXPORT WINNER
The XK120 was a great success as an export earner, with over 85 per cent of all XK120s going to foreign climes.

WHEELS
Standard wheels were the same steel discs as on the Jaguar saloons.

SPECIFICATIONS

MODEL Jaguar XK120 (1949–54)

PRODUCTION 12,055

BODY STYLES Two-seater roadster, fixed-head coupé, and drophead coupé.

CONSTRUCTION Separate chassis, aluminium/steel bodywork.

ENGINE 3442cc twin overhead cam six-cylinder, twin SU carburettors.

POWER OUTPUT 160 bhp at 5100 rpm.

TRANSMISSION Four-speed manual, Moss gearbox with syncromesh on upper three ratios.

SUSPENSION *Front:* independent, wishbones and torsion bars; *Rear:* live rear axle, semi-elliptic.

BRAKES Hydraulically operated 30-cm (12-in) drums.

MAXIMUM SPEED 203 km/h (126 mph)

0–60 MPH (0–96 KM/H) 10 sec

0–100 MPH (0–161 KM/H) 35.3 sec

A.F.C. 6.1–7.8 km/l (17–22 mpg)

ROADSTER REVIVAL
Even though numbers of roadsters
were trimmed further in the late
Eighties' scrabble to restore them, their
flowing curves and perfect proportions
are now more widely appreciated.

LIMITED VISION
*Fixed-head coupés had
limited rear vision, but
at least you stayed dry
in a British summer.*

MIDAS TOUCH
With the XK120, once again
Jaguar Boss William Lyons had
pulled off his favourite trick:
offering sensational value for
money compared with anything
else in its class. In fact this time
there was nothing else in its class.

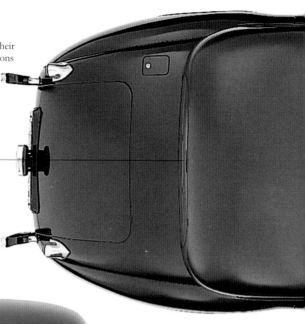

INTERIOR
Surrounded by leather and thick pile
carpet, you feel good just sitting in an
XK120 – a lush interior, purposeful
instruments, and the bark of the exhaust

SELLING THE DREAM
Beautiful enough as it was, the original sales brochure for the XK120 used airbrushed photographs of the very first car built – the 1948 Motor Show car.

ENGINE
The famed XK six-cylinder engine was designed by Bill Heynes and Wally Hassan, and went on to power the E-Type *(see pages 306–09)* and other Jaguars up until 1986. Even this was "styled"; William Lyons insisted it had twin camshafts to make it resemble GP cars of the Thirties.

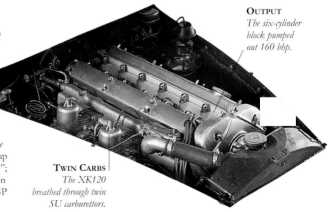

OUTPUT
The six-cylinder block pumped out 160 bhp.

TWIN CARBS
The XK120 breathed through twin SU carburettors.

JAGUAR *C-Type*

THE C-TYPE IS THE CAR that launched the Jaguar racing legend and began a Le Mans love affair for the men from Coventry. In the 1950s, Jaguar boss Bill Lyons was intent on winning Le Mans laurels for Britain, just as Bentley had done a quarter of a century before. After testing mildly modified XK120s in 1950, Jaguar came up with a competition version, the XK120C (C-Type) for 1951. A C-Type won that year, failed in 1952, then won again in 1953. By then the C-Type's place in history was assured, for it had laid the cornerstone of the Jaguar sporting legend that blossomed through its successor, the D-Type, which bagged three Le Mans 24-hour wins in four years. C-Types were sold to private customers, most of whom used them for racing rather than road use. They were tractable road cars though, often driven to and from meetings; after their days as competitive racers were over, many were used as high-performance highway tourers.

PRODUCTION-CAR LINK
Jaguar's Bill Lyons dictated that the C-Type racer should bear a strong family resemblance to production Jaguars, and the Malcolm Sayer body, fitted to a special frame, achieved that aim.

FAST FUELLING
Quick-release filler-cap was another racing feature, and could save valuable seconds in a race.

ACCESSIBILITY
It was easier to step over the door than open it; the passenger did not even get one.

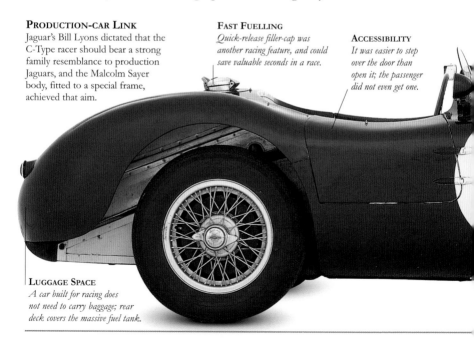

LUGGAGE SPACE
A car built for racing does not need to carry baggage; rear deck covers the massive fuel tank.

NICE AMALGAM

The clever blend of beauty and function retained the pouncing-cat Jaguar "look", while creating an aerodynamically efficient tool for the high-speed Le Mans circuit.

BRAKES
The C-Type introduced disc brakes to road racing in 1952, though most examples used drums.

SPECIFICATIONS

MODEL Jaguar C-Type (1951–53)

PRODUCTION 53

BODY STYLE Two-door, two-seater sports racer.

CONSTRUCTION Tubular chassis, aluminium body.

ENGINE Jaguar XK120 3442cc, six-cylinder, double overhead camshaft with twin SU carburettors.

POWER OUTPUT 200–210 bhp at 5800 rpm.

TRANSMISSION Four-speed XK gearbox with close-ratio gears.

SUSPENSION Torsion-bars all round; wishbones at front, rigid axle at rear.

BRAKES Lockheed hydraulic drums; later cars used Dunlop discs all round.

MAXIMUM SPEED 232 km/h (144 mph)

0–60 MPH (0–96 KM/H) 8.1 sec

0–100 MPH (0–161 KM/H) 20.1 sec

A.F.C. 5.7 km/l (16 mpg)

SUSPENSION
Telescopic dampers smoothed the ride.

RACE MODELS
The C-Type was always most at home on the track, though more at Le Mans – where it won the 24-hour classic twice from three attempts – than on shorter circuits such as Silverstone.

ENGINE
The engine was taken from the XK120 and placed into the competition version. Horsepower of the silky six was boosted each year until some 220 bhp was available.

BONNET
Bonnet hinged forwards to ease mid-race adjustments.

BLOCK POSITION
Engine snuggled neatly into its bay, ready for action.

AERO INFLUENCE
Designer Malcolm Sayer's aircraft industry background shows through in the smooth aerodynamic styling. Louvres on the bonnet help hot air escape; the engine cover is secured by quick-release handles and leather safety straps.

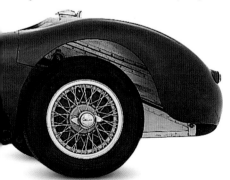

INTERIOR
The cockpit was designed for business, not comfort, but was roomy enough for two adults; passengers were provided with a grab-handle in case the driver thought he was at Le Mans. In racing trim, cars ran with a single aero-screen; this car has an additional full-width screen.

JAGUAR *XK150*

THE XK150 APPEARED IN the Spring of 1957 and was the most refined of the XK trio. One of the last Jaguars to have a separate chassis, the 150 marked the beginning of the civilization of the Jaguar sports car. With its wider girth and creature comforts, it was to hold the market's interest until the then-secret E-Type project *(see pages 306–09)* was ready for unveiling in 1961. In the late 1950s, the XK150 was a seriously glamorous machine, almost as swish as an Aston Martin, but £1,500 cheaper. March 1958 saw more power with the "S" performance package, which brought the 3.4 up to 250 bhp, and in 1959 the 3.8's output soared to 265 bhp. Available as a roadster, drophead, or fixed-head coupé, the 150 sold a creditable 9,400 examples in its four-year run. Despite being eclipsed by the E-Type, the 150 was charismatic enough to be the personal transport for racing ace Mike Hawthorn and startlet Anita Ekberg.

SALOON REAR
From the rear, the fixed-head has definite saloon lines, with its curved rear screen, big wrap-around bumper, wide track, and cavernous boot.

REDUCED PRICE
XK150s have fallen in price and can now be bought for the same price as an Austin Healey 3000 *(see pages 52–55)*, a Daimler Dart *(see pages 190–93)*, or a Sunbeam Tiger *(see pages 438–39)*.

RARE COUPÉ
*The rarest model is the
XK150S drophead coupé,
with a mere 193 cars built.*

CLASSIC STRAIGHT-SIX
*The classic, twin overhead-cam
design first saw the light of
day in 1949, and was phased
out as recently as 1986.*

PURE CAT PROFILE
The gorgeous curved body sits on a
conventional chassis. Joints and curves
were smoothed off at the factory
using lead. The 1950s' motor industry
paid little thought to rustproofing, so
all Jaguars of the period are shameful
rust-raisers.

SPECIFICATIONS

MODEL Jaguar XK150 FHC (1957–61)

PRODUCTION 9,400

BODY STYLES Two-seater roadster,
drophead, or fixed-head coupé.

CONSTRUCTION Separate pressed-steel
chassis frame with box section
side members.

ENGINES Straight-six, twin overhead-cam
3442cc or 3781cc.

POWER OUTPUT 190 bhp at 5500 rpm
(3.4); 210 bhp at 5500 rpm (3.8);
265 bhp at 5500 rpm (3.8S)

TRANSMISSION Four-speed manual, with
optional overdrive, or three-speed Borg
Warner Model 8 automatic.

SUSPENSION Independent front, rear leaf
springs with live rear axle.

BRAKES Dunlop front and rear discs.

MAXIMUM SPEED 217 km/h (135 mph)

0–60 MPH (0–96 KM/H) 7.6 sec (3.8S)

0–100 MPH (0–161 KM/H) 18 sec

A.F.C. 6.4 km/l (18 mpg)

JAGUAR *E-Type*

WHEN JAGUAR BOSS WILLIAM LYONS, by now Sir William, unveiled the E-Type Jaguar at the Geneva Motor Show in March 1961, its ecstatic reception rekindled memories of the 1948 British launch of the XK120 *(see pages 296–99)*. The E-Type, or XKE as it is known in America, created a sensation. British motoring magazines had produced road tests of pre-production models to coincide with the launch – and yes, the fixed-head coupé really could do 242 km/h (150.4 mph). OK, so the road-test cars were perhaps tweaked a little and early owners found 233 km/h (145 mph) a more realistic maximum, but the legend was born. It was not just a stunning, svelte sports car though; it was a trademark Jaguar sporting package, once again marrying sensational performance with superb value for money. Astons and Ferraris, for example, were more than double the price.

BEST-OF-BREED
The impact the shape made at its launch on 15 March 1961 at the Geneva Motor Show, is now the stuff of Jaguar lore. Those first E-Type roadsters and fixed-head coupés, produced until June 1962, are now referred to as "flat-floor" models and they are the most prized of all. In fact, their flat floor was something of a flaw, as recessed foot wells were later incorporated to increase comfort for taller drivers.

HANDLING
Jaguar designed an all-new independent set up at the rear. Handling in the wet and on the limit is often criticized, but for its day the E-Type was immensely capable.

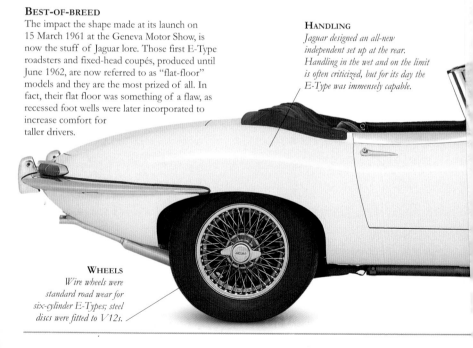

WHEELS
Wire wheels were standard road wear for six-cylinder E-Types; steel discs were fitted to V12s.

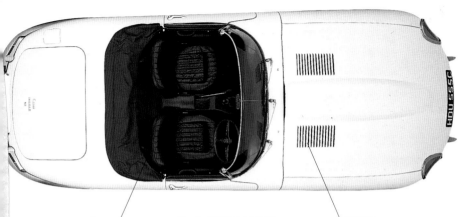

SEATS
Thin-backed bucket seats of the 3.8s were criticized. In the 4.2, as here, they were greatly improved.

SIMPLICITY OF LINE
Designer Malcolm Sayer insisted he was an aerodynamicist and hated to be called a stylist. He claimed the E-Type was the first production car to be "mathematically" designed.

VENTS
Louvres are not for looks; E-Types, particularly early ones, tended to overheat in hot climates.

LENS COVERS
The stylish but inefficient lens covers were removed in 1967.

BRAKES
All-round disc brakes as standard were part of the spec from first E-Types.

HOOD
Hood was neatly tucked away beneath a fitted tonneau cover.

BUMPERS
Chromed slimline bumpers were beautiful but offered no protection.

SPECIFICATIONS

MODEL E-Type Jaguar (1961–74)

PRODUCTION 72,520

BODY STYLES Two-seater roadster and fixed coupé, 2+2 fixed head coupé.

CONSTRUCTION Steel monocoque.

ENGINES 3781cc straight-six; 4235cc straight-six; 5343cc V12.

POWER OUTPUT 265 to 272 bhp.

TRANSMISSION Four-speed manual, optional automatic from 1966.

SUSPENSION *Front:* independent, wishbones and torsion bar; *Rear:* independent, coil and radius arm.

BRAKES Discs all round.

MAXIMUM SPEED 241 km/h (150 mph) (3.8 & 4.2); 230 km/h (143 mph) (5.3)

0–60 MPH (0–96 KM/H) 7–7.2 sec

0–100 MPH (0–161 KM/H) 16.2 sec (3.8)

A.F.C. 5.7–7 km/l (16–20 mpg)

TELL TAIL

The thin bumpers with lights above are an easy giveaway for E-Type spotters. From 1968, with the introduction of the Series 2, bulkier lamp clusters appeared below the bumpers. A detachable hardtop was available as an option.

US MARKET

The E-Type's amazing export success is summed up by the fact that of every three built, two were exported. Fixed-head coupés actually accounted for a little over half of all E-Type production, yet the roadster was the major export winner, with most going to the US. Ironically, though, it was American emission regulations that were increasingly strangling the Cat's performance.

WIPERS
Unusual and sporty-looking triple wipers gave way to a two-blade system with the 1971 V12.

INTERIOR
The interior of this Series 1 4.2 is the epitome of sporting luxury, with leather seats, wood-rim wheel, and an array of instruments and toggle switches – later replaced by less sporting rocker and less injurious rocker switches. The 3.8s had an aluminium-finished centre console panel and transmission tunnel.

CLASSY BONNET
This view of the E-Type's bulging, sculptured bonnet is still the best of any car.

HDU 555C

JENSEN *Interceptor*

THE JENSEN INTERCEPTOR WAS one of those great cars that comes along every decade or so. Built in a small Birmingham factory, a triumph of tenacity over resources, the Interceptor's lantern-jawed looks and tyre-smoking power made the tiny Jensen company a household name. A glamorous cocktail of an Italian-styled body, American V8 engine, and genteel British craftsmanship, it became the car for successful swingers of the late 1960s and 1970s. The Interceptor was handsome, fashionable, and formidably fast, but its tragic flaw was a single figure appetite for fuel – 3.5 km/l (10 mpg) if you enjoyed yourself. After driving straight into two oil crises and a worldwide recession, as well as suffering serious losses from the ill-starred Jensen-Healey project, Jensen filed for bankruptcy in 1975 and finally closed its doors in May 1976.

TIMELESS STYLING
The Interceptor's futuristic shape hardly changed over its 10-year life span and was widely acknowledged to be one of the most innovative designs of its decade. The classic shape was crafted by Italian styling house Vignale. From bare designs to running prototype took just three months.

SCREEN
Rear screen lifted up to reveal a large luggage compartment.

BODYWORK
Bodies were all-steel, with little attention paid to corrosion proofing. Early cars were tragic rust-raisers.

SPECIFICATIONS

MODEL Jensen Interceptor (1966–76)

PRODUCTION 1,500

BODY STYLE All-steel occasional four-seater coupé.

CONSTRUCTION Separate tubular and platform type pressed steel frame.

ENGINE 6276cc V8.

POWER OUTPUT 325 bhp at 4600 rpm.

TRANSMISSION Three-speed Chrysler TorqueFlite automatic.

SUSPENSION Independent front with live rear axle.

BRAKES Four-wheel Girling discs.

MAXIMUM SPEED 217 km/h (135 mph)

0–60 MPH (0–96 KM/H) 7.3 sec

0–100 MPH (0–161 KM/H) 19 sec

A.F.C. 4.9 km/l (13.6 mpg)

BEAUTIFUL INTERIOR
Road testers complained that the Interceptor's dash was like the flight deck of a small aircraft, but the interior was beautifully hand-made with the finest hides and plush Wilton carpets.

SIMPLE MECHANICS
With one huge carburettor and only a single camshaft, the Interceptor had a simple soul.

ENGINE
The lazy Chrysler V8 of 6.2 litres gave drag-strip acceleration and endless reliability.

TYRES
Dunlop SPs replaced skinny pre-'68 RS5s.

KAISER *Darrin*

"THE SPORTS CAR THE WORLD has been awaiting" was a monster flop. Designed by Howard "Dutch" Darrin, Kaiser's odd hybrid came about in 1953 as an accident. Henry J. Kaiser, the ill-mannered chairman of the Kaiser Corporation, had so riled Darrin that he disappeared to his California studio, spent his own money, and created a purse-lipped two-seater that looked like it wanted to give you a kiss. Its futuristic glass-fibre body rode on a Henry J. chassis and was powered by a Willys six-pot mill. Alas, the body rippled and cracked, the sliding doors wouldn't slide, and the weedy 90 bhp flathead was no match for Chevy's glam Corvette. At a costly $3,668, the Darrin was in Cadillac territory, and only 435 found buyers. Late in '54, Kaiser-Willys went under, taking the Darrin with them. Few mourned either's demise.

A TRUE CLASSIC
The Darrin was beautifully styled and, unlike most visions of the future, has hardly dated at all. The Landau top could be removed and a hardtop fitted, and, with its three-speed floor shift and overdrive, it could return up to a remarkable 9.6 km/l (27 mpg).

BODY SHELL
Darrin bodies were made by boat-builders Glasspar.

WING SHAPE
Rear wing tapers upwards to create a fine torpedo-like shape.

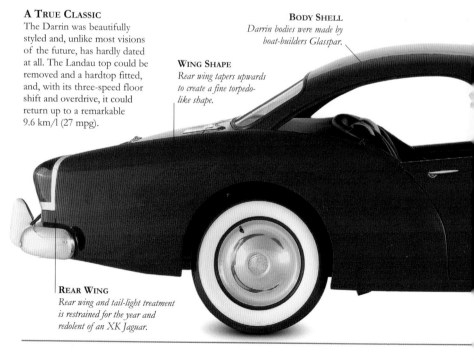

REAR WING
Rear wing and tail-light treatment is restrained for the year and redolent of an XK Jaguar.

INTERIOR

Standard equipment included electric wipers, tachometer, and a European-style dashboard, with leather trim an optional extra. Whitewall tyres and a one-piece windscreen were also standard.

SPECIFICATIONS

MODEL Kaiser Darrin 161 (1954)

PRODUCTION 435 (total)

BODY STYLE Two-seater sports.

CONSTRUCTION Glass-fibre body, steel frame.

ENGINE 161cid six.

POWER OUTPUT 90 bhp.

TRANSMISSION Three-speed manual with optional overdrive.

SUSPENSION *Front:* coil springs; *Rear:* leaf springs.

BRAKES Front and rear drums.

MAXIMUM SPEED 161 km/h (100 mph)

0–60 MPH (0–96 KM/H) 15.1 sec

A.F.C. 9.6 km/l (27 mpg)

SLIDING DOORS

Howard Darrin first conceived his contentious sliding doors back in 1922. The trouble was that they rattled, jammed, and didn't open all the way.

LIMITED SIX

The six-cylinder unit produced a top speed of only 161 km/h (100 mph).

SLEEK SELL

Adverts called it "the outstanding pleasure car of our day".

LATE DELIVERIES
The Darrin took its time coming. It was first announced on 26 September 1952, with 60 initial prototypes eventually displayed to the public on 11 February 1953. Final production cars reached owners as late as 6 January 1954.

RISING ARCH
Undeniably pretty, the wing line slopes down through the door and meets a dramatic kick-up over the rear wheelarch.

CABIN SPACE
Hardtop made the cabin much less claustrophobic and cramped than that of the soft-top model.

SIDESCREEN

Swivelling Plexiglas sidescreens reduced cockpit buffeting.

BELT UP
The Darrin was remarkable for being only the third US production car to feature seat belts as standard. The other two cars were a Muntz and a Nash.

HEADLIGHTS

The prototype headligh¡ height was too low for state lighting laws, so Kaiser stylists hiked up the front wing line for the real thing.

CHASSIS
*Stock Henry J. chassis
and engine didn't do much
for the Darrin's bloodline.*

ENGINE
Kaiser opted for an F-head Willys version of the
Henry J. six-pot motor but, with just one carb,
it boasted only 10 more horses than standard.
After the company folded, Darrin dropped
300 bhp supercharged Caddy V8s into the
remaining cars, which went like hell.

VW-STYLE
*Front aspect looks very
much like an early
VW Karmann Ghia.*

PRICING
*The 90 bhp Darrin
cost $145 more than the
150 bhp Chevy Corvette.*

AN UNHAPPY ALLIANCE
Henry J. Kaiser was livid that
Howard Darrin had worked on
the car without his permission.
In the end, the Darrin was
actually saved by Henry J.'s
wife, who reckoned it
was "the most beautiful
thing" she'd ever seen.

BRIT REAR
*Rear aspect is
surprisingly British-
looking for a
Californian design.*

KAISER *Henry J. Corsair*

IN THE EARLY 1950s, the major motor manufacturers reckoned that small cars meant small profits, so low-priced transportation was left to independent companies like Nash, Willys, and Kaiser-Frazer. In 1951, a streamlined, Frazer-less Kaiser launched "America's Most Important New Car", the Henry J. An 80 bhp six-cylinder "Supersonic" engine gave the Corsair frugal fuel consumption, with Kaiser claiming that every third mile in a Henry J. was free. The market, however, was unconvinced. At $1,561, the Corsair cost more than the cheapest big Chevy, wasn't built as well, and depreciated rapidly. Small wonder then that only 107,000 were made. Had America's first serious economy car been launched seven years later during the '58 recession, the Henry J. may well have been a best-seller.

SMALLER MODEL
The stubborn head of Kaiser industries insisted that the Henry J., originally designed as a full-size car by designer Howard "Dutch" Darrin, be scaled down.

WINGS
Bolt-on front and rear wings were part of the Henry J.'s money-saving philosophy.

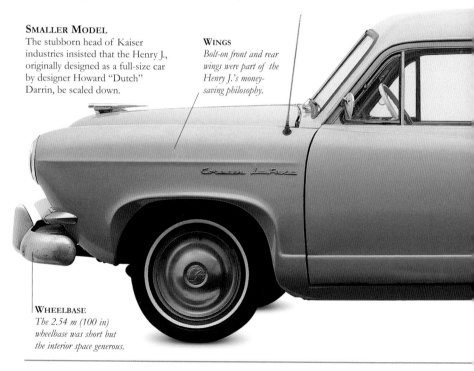

WHEELBASE
The 2.54 m (100 in) wheelbase was short but the interior space generous.

DASH CONTROLS
The few controls included starter, ignition, light, and choke switches.

SPECIFICATIONS

MODEL Kaiser Henry J. Corsair Deluxe (1952)

PRODUCTION 12,900 (1952)

BODY STYLE Two-door, five-seater sedan.

CONSTRUCTION Steel body and chassis.

ENGINES 134cid four, 161cid six.

POWER OUTPUT 68–80 bhp.

TRANSMISSION Three-speed manual with optional overdrive, optional three-speed Hydra-Matic automatic.

SUSPENSION *Front:* coil springs; *Rear:* leaf springs with live axle.

BRAKES Front and rear drums.

MAXIMUM SPEED 140 km/h (87 mph)

0–60 MPH (0–96 KM/H) 17 sec

A.F.C. 12 km/l (34 mpg)

INSIDE THE CORSAIR
The interior was seriously austere. Apart from overdrive and auto transmission, very few options were available.

ROOF LINE
High roof line owed its existence to the fact that Kaiser's chairman always wore a hat.

COLOURS
Blue Satin was one of nine colours available.

BOOT SPACE
With the rear seat folded down, the luggage space was among the largest of any passenger sedan.

LAMBORGHINI *Miura*

THE LAUNCH OF THE LAMBORGHINI MIURA at the 1966 Geneva Motor Show was the decade's motoring sensation. Staggeringly beautiful, technically pre-eminent, and unbelievably quick, it was created by a triumvirate of engineering wizards all in their twenties. For the greater part of its production life the Miura was reckoned to be the most desirable car money could buy, combining drop-dead looks, awesome performance, and unerring stability, as well as an emotive top speed of 282 km/h (175 mph). From its dramatic swooping lines – even Lamborghini thought it was too futuristic to sell – to its outrageously exotic colours, the Miura perfectly mirrored the middle Sixties. But, as the oil crises of the Seventies took hold, the Miura slipped into obscurity, replaced in 1973 by the unlovely, and some say inferior, Countach *(see pages 322–25).*

GT40 LINKS

In looks and layout the mid-engined Lambo owes much to the Ford GT40 *(see pages 258–61)* but was engineered by Gianpaolo Dallara. At the core of the Miura is a steel platform chassis frame with outriggers front and rear to support the major mechanicals. The last-of-the-line SV was the most refined Miura, with more power, a stiffer chassis, and redesigned suspension.

INSULATION

In an attempt to silence a violently loud engine, Lamborghini put 10 cm (4 in) of polystyrene insulation between engine and cabin.

REAR WING

The rear wing profile was different on the SV than on earlier models.

LIGHTS
Standard Miura headlights were shared by the Fiat 850.

HEIGHT
The Miura only came up to waist height – just 107 cm (42 in).

INTERIOR
Standard interior trim was unimpressive oatmeal vinyl.

ENDURING STYLE
Long, low, and delicate, the Miura is still considered one of the most handsome automotive sculptures ever. The car was so low that headlights had to be "pop-up" to raise them high enough for adequate vision.

RARE SV
Only 150 SVs were built. Very few had a "split sump" that had separate oil for the engine and gearbox.

ACCELERATION
Acceleration still compares well with modern supercars.

NEW JERSEY
FLA 352
GARDEN STATE

TAIL-END ACTION

Because the Miura sits so low, it displays virtually zero body roll; therefore there is little warning before the tail breaks away, which, with all that power, is likely to happen at high speeds.

ENGINE

The V12 4-litre engine was mid-mounted transversely to prevent the car's wheelbase from being too long. The gearbox, final drive and crankcase were all cast in one piece to save space. Beneath the pipery slumber 12 pistons, four chain-driven camshafts, 24 valves, and four carburettors.

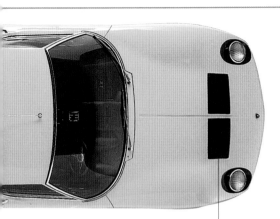

SPECIFICATIONS

MODEL Lamborghini Miura (1966–72)

PRODUCTION Approx 800.

BODY STYLE Two-seater roadster.

CONSTRUCTION Steel platform chassis, light alloy and steel bodywork.

ENGINE Transverse V12 4.0 litre.

POWER OUTPUT P400, 350 bhp at 7000 rpm; P400S, 370 bhp at 7700 rpm; P400SV, 385 bhp at 7850 rpm.

TRANSMISSION Five-speed with trans axle.

SUSPENSION Independent front and rear.

BRAKES Four-wheel ventilated disc.

MAXIMUM SPEED 282 km/h (175 mph) (P400SV)

0–60 MPH (0–96 KM/H) 6.7 sec

0–100 MPH (0–161 KM/H) 15.1 sec

A.F.C. 5.7 km/l (16 mpg)

LIGHT POWERHOUSE
The Miura has a very impressive power-to-weight ratio – it's able to produce 385 bhp, yet it weighs only 1,200 kg (2,646 lb).

FILLER CAP
The petrol filler hid under the bonnet slat.

FRONT LIFT
Treacherous aerodynamics meant that approaching speeds of 274 km/h (170 mph) both of the Miura's front wheels could actually lift off the ground.

GEARBOX
The gearbox was a disappointment, with a truck-like, sticky action that did not do the Miura's gorgeous engine any justice.

INTERIOR
The cockpit is basic but finely detailed, with a huge Jaeger speedo and tacho. Six minor gauges on the left of the console tell the mechanical story. The alloy gear-lever gate is a hand-made work of art.

LAMBORGHINI *Countach 5000S*

THE COUNTACH WAS FIRST UNVEILED at the 1971 Geneva Motor Show as the Miura's replacement, engineered by Giampaolo Dallara and breathtakingly styled by Marcello Gandini of Bertone fame. For a complicated, hand-built car, the Countach delivered all the reliable high performance that its swooping looks promised. In 1982, a 4.75-litre 375 bhp V12 was shoe-horned in to give the up-coming Ferrari Testarossa *(see pages 244–47)* something to reckon with. There is no mid-engined car like the Countach. The engine sits longitudinally in a multi-tubular space frame, with fuel and water carried by twin side-mounted tanks and radiators. On the down side, visibility is appalling, steering is heavy, gear selection recalcitrant, and the cockpit is cramped. Yet such faults can only be considered as charming idiosyncrasies when set against the Countach's staggering performance – a howling 301 km/h (187 mph) top speed and a 0–60 (96 km/h) belt of 5 seconds.

BREAKING THE RULES
The shape is a riot of creative genius that ignores all established rules of car design. Air scoops under the body's side windows break up the wedge-shaped line and form a ready-made indent for a compact door catch.

AIR SCOOPS
Air scoops provided ideal hand-holds for the huge scissor doors.

SOUND EFFECTS
Inches away, all occupants were able to hear exactly what this engine had to say.

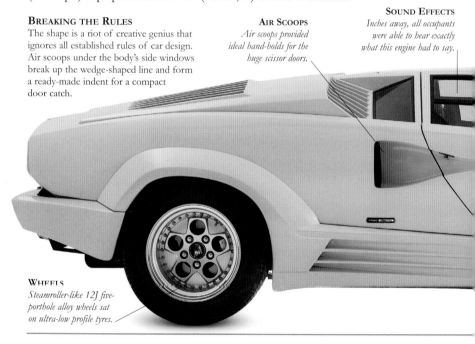

WHEELS
Steamroller-like 12J five-porthole alloy wheels sat on ultra-low profile tyres.

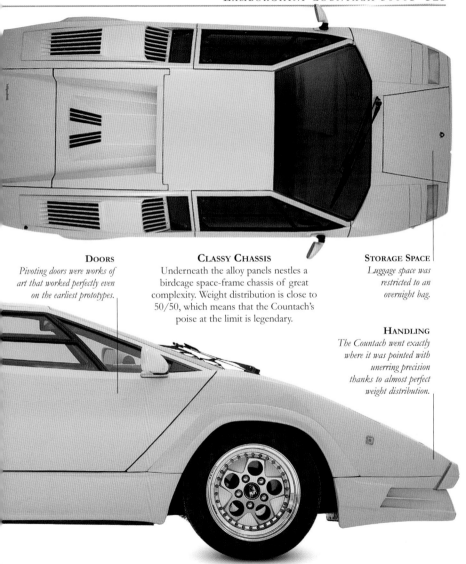

DOORS
Pivoting doors were works of art that worked perfectly even on the earliest prototypes.

CLASSY CHASSIS
Underneath the alloy panels nestles a birdcage space-frame chassis of great complexity. Weight distribution is close to 50/50, which means that the Countach's poise at the limit is legendary.

STORAGE SPACE
Luggage space was restricted to an overnight bag.

HANDLING
The Countach went exactly where it was pointed with unerring precision thanks to almost perfect weight distribution.

INTERIOR
The cabin was crude, with unsubtle interior architecture. Switches and wands were Fiat- and Lancia-sourced. Scant body protection means that most Countachs acquire a tapestry of scars.

SUSPENSION
Independent front and rear suspension had double wishbones and coil springs.

CELEBRATIONS
The 25-year anniversary of Lamborghini production in 1985 was celebrated with the 5000S and the elite Quattrovalvole 5000S.

SPECIFICATIONS

MODEL Lamborghini Countach (1973–90)

PRODUCTION Approx 1,000

BODY STYLE Mid-engined, two-seater sports coupé.

CONSTRUCTION Alloy body, space-frame chassis.

ENGINE 4754cc four-cam V12.

POWER OUTPUT 375 bhp at 7000 rpm.

TRANSMISSION Five-speed manual.

SUSPENSION Independent front and rear with double wishbones and coil springs.

BRAKES Four-wheel vented discs.

MAXIMUM SPEED 301 km/h (187 mph)

0–60 MPH (0–96 KM/H) 5.1 sec

0–100 MPH (0–161 KM/H) 13.3 sec

A.F.C. 3.2 km/l (9 mpg)

GRAND AUTO

Everything on the Countach
s built on a grand scale.
Four exhausts,
four camshafts,
12 cylinders, half
a dozen 45DCOE
Webers, and
the widest track
of any car on
the road.

AMPLE GIRTH
*It took an epoch to get used
to the extra wide body.*

MANOEUVRABILITY
Reversing the Countach is a bit
like launching the *Queen Mary*. The
preferred technique is to open the
scissor door and sit on the sill while
looking over your shoulder.

ECONOMY?
*The 4.75-litre power unit spared no
thought for fuel economy and drank
one litre of fuel every 3.2 km (9 mpg).*

LANCIA *Aurelia B24 Spider*

BEAUTY IS MORE THAN JUST skin deep on this lovely little Lancia, for underneath those lean Pininfarina loins the Aurelia's innards bristle with innovative engineering. For a start there is the compact alloy V6. Designed under Vittorio Jano, the man responsible for the great racing Alfas of the Twenties and Thirties, this free-revving, torquey little lump was the first mass-produced V6. The revolution was not just at the front though, for at the back were the clutch and gearbox, housed in the transaxle to endow the Aurelia with near-perfect weight distribution. These innovations were first mated with the Pininfarina body in 1951, producing the Aurelia B20 GT coupé, often credited as the first of the new breed of modern post-war GTs. And the point of it all becomes clear when you climb behind the wheel, for although the Aurelia was never the most accelerative machine, its handling was so impeccable that 40 years on it still impresses with its masterly cornering poise.

FAMILY RESEMBLANCE

The Spider bears a passing family resemblance to the Aurelia saloon, and even more so to the GT models. Neither of the closed versions had the wrap-around windscreen though, or the equally distinctive half-bumpers; the Spider's radiator grille was a slightly different shape, too.

LUGGAGE ROOM

The Aurelia Spider scored well in luggage-carrying capabilities compared with other two-seaters of the time.

DESIGNATIONS

The Spider and convertible were designated B24 Aurelias; B10, 15, 21, and 22 were four-door saloons, and B20 the GT coupé.

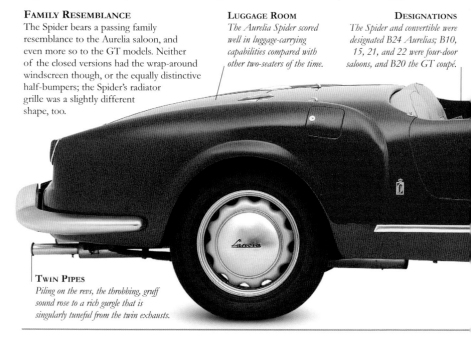

TWIN PIPES

Piling on the revs, the throbbing, gruff sound rose to a rich gurgle that is singularly tuneful from the twin exhausts.

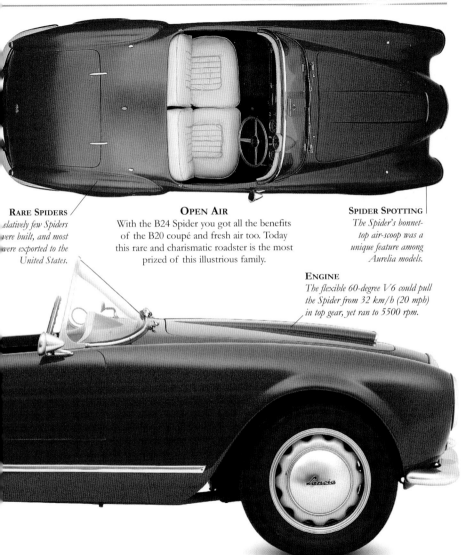

RARE SPIDERS
elatively few Spiders
were built, and most
were exported to the
United States.

OPEN AIR
With the B24 Spider you got all the benefits
of the B20 coupé and fresh air too. Today
this rare and charismatic roadster is the most
prized of this illustrious family.

SPIDER SPOTTING
The Spider's bonnet-
top air-scoop was a
unique feature among
Aurelia models.

ENGINE
The flexible 60-degree V6 could pull
the Spider from 32 km/h (20 mph)
in top gear, yet ran to 5500 rpm.

RIGHT-HAND DRIVE
Until the Aurelia, Lancia had eccentrically persisted with right-hand steering, even for the home market. The adoption of left-hand drive makes this right-hooker a real rarity.

INTERIOR
The panel has just three major dials and a clutch of switches on a painted metal dash. It was devoid of the walnut-leather trimmings which British car makers of the time considered essential for a luxury sports car. The elegant, adjustable Nardi steering wheel was standard equipment on the Spider.

ENGINE
Aurelias featured the world's first mass-production V6, an all alloy unit which progressively grew from 1754cc to 1991cc, to the 2451cc fitted to the B24 Spider.

BALANCE
For perfect balance, the weight of the engine was offset by locating clutch and gearbox in a unit with differential at the rear.

SWEEPING WINGS

The curvaceous Pininfarina shape is characterized by the sweeping front wings and long luggage compartment. The Spider's high-silled monocoque construction meant that the doors were small. Protection from the elements was fairly basic; the B24 had a simple hood with plastic sidescreens.

FLAG BADGES

These represent the joint input of Lancia, designers and manufacturers of the mechanical parts, and Pininfarina, who styled the body and built the cars.

RACE PEDIGREE

The B20 GT coupés, from which the B24 Spider was derived, achieved a second overall on the Mille Miglia and a Le Mans class win.

SPECIFICATIONS

MODEL Lancia Aurelia B24 Spider (1954–56)

PRODUCTION 330

BODY STYLE Two-seater sports convertible.

CONSTRUCTION Monocoque with pressed steel and box-section chassis frame.

ENGINE Twin-overhead-valve aluminium alloy V6, 2451cc.

POWER OUTPUT 118 bhp at 5000 rpm.

TRANSMISSION Four-speed manual.

SUSPENSION Sliding pillar with beam axle and coil springs at front, De Dion rear axle on leaf springs.

BRAKES Hydraulic, finned alloy drums, inboard at rear.

MAXIMUM SPEED 180 km/h (112 mph)

0–60 MPH (0–96 KM/H) 14.3 sec

A.F.C. 7.8 km/l (22 mpg)

LANCIA *Stratos*

THE LANCIA STRATOS WAS BUILT as a rally-winner first and a road car second. Fiat-owned Lancia took the bold step of designing an all-new car solely to win the World Rally Championship and, with a V6 Ferrari Dino engine *(see pages 234–37)* on board, the Stratos had success in 1974, '75, and '76. Rallying rules demanded that at least 500 cars be built, but Lancia needed only 40 for its rally programme; the rest lay unsold in showrooms across Europe for years and were even given away as prizes to high-selling Lancia dealers. Never a commercial proposition, the Stratos was an amazing mix of elegance, hard-charging performance, and thrill-a-minute handling.

STUBBY STYLE

Shorter than a Mk II Escort, and with the wheelbase of a Fiat 850, the stubby Stratos wedge looks almost as wide as it is long. The front and back of the car are glass-fibre with a steel centre-section. The constant radius windscreen is cut from a cylindrical section of thin glass to avoid distortion. Whatever the views on the Stratos' styling, though, there is no doubting the fact that the glorious metallic soundtrack is wonderful.

WHEELS
Campagnallo alloys sat on Pirelli P7F rubber – F stands for a soft compound to give a gentler loss of adhesion.

ASSEMBLY
Bertone built the bodies, while Lancia added their sometimes-clumsy finishing touches at the Chivasso factory in Turin.

SHARP END
Flimsy nose section concealed spare wheel, radiator, and twin thermostatically controlled cooling fans.

REAR ASPECT

A 1970s fad, the matt black plastic rear window slats did little for rearward visibility. The raised rear spoiler did its best to keep the rear wheels stuck to the road like lipstick on a collar.

SPECIFICATIONS

MODEL Lancia Stratos (1973–80)

PRODUCTION 492

BODY STYLE Two-seater mid-engined sports coupé.

CONSTRUCTION Glass-fibre and steel unit construction body chassis tub.

ENGINE 2418cc mid-mounted transverse V6.

POWER OUTPUT 190 bhp at 7000 rpm.

TRANSMISSION Five-speed manual in unit with engine and transaxle.

SUSPENSION Independent front and rear with coil springs and wishbones.

BRAKES Four-wheel discs.

MAXIMUM SPEED 230 km/h (143 mph)

0–60 MPH (0–96 KM/H) 6.0 sec

0–100 MPH (0–161 KM/H) 16.7 sec

A.F.C. 6.4 km/l (18 mpg)

SAFETY BAR
Safety bar was to protect the cabin in case the car rolled.

SUSPENSION
Rear springing was by Lancia Beta-style struts, with lower wishbones.

INTERIOR

The Stratos was hopeless as a day-to-day machine, with a claustrophobic cockpit and woeful rear vision. The width of 1.72 m (67 in) and the narrow cabin meant that the steering wheel was virtually in the middle of the car. Quality control was dire, with huge panel gaps, mischievous electrics, and ventilation that did not work.

COMFORT
Truncated cabin was cramped, cheap, nasty, and impossibly hot.

RACE UNIT
Factory rally versions had a four-valve V6 engine.

WEIGHT
The Stratos was a two-thirds glass-fibre featherweight, tipping the scales at a whisker over 908 kg (2,000 lb).

A DRIVER'S CAR

Driving a Stratos hard isn't easy. You sit almost in the middle of the car with the pedals offset to the left and the steering wheel to the right. Ferocious acceleration, monumental oversteer, and lots of heat from the engine make the Stratos a real handful.

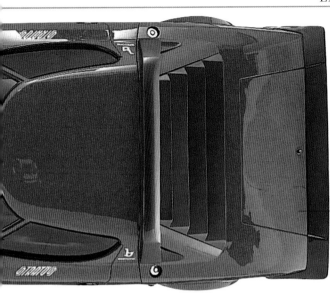

RALLY SUCCESS
Lancia commissioned Bertone to build a "take-no-prisoners" rally weapon, and the Stratos debuted at the 1971 Turin Show. Despite scooping three World Championships, sales of Stratos road cars were so slow that they were still available new up until 1980.

REAR COWL
Moulded glass-fibre rear cowl lifted up by undoing two clips, giving access to midships-mounted power plant.

ENGINE
Lifted straight out of the Dino 246, the 190 bhp transverse, mid-mounted V6 has four chain-driven camshafts spinning in alloy heads, which sit just 15 cm (6 in) from your ear. Clutch and throttle are incredibly stiff, which makes smooth driving an art form.

DEEP WINDOWS
Perspex side windows are so deeply recessed within the bodywork that they can be fully opened without causing any wind turbulence.

LEXUS *LFA*

FOR THEIR 20TH BIRTHDAY LEXUS went crazy and built a supercar. But what a car! The LFA is like no other Lexus, with F1 technology, 65 per cent carbon-fibre construction, 322 km/h (200 mph) top speed, and one of the best engine notes in the world. The 4.8-litre V10 is so fast-spinning that it can rev from idle to 9,000 rpm in 0.6 seconds – too fast even for a conventional tachometer. The six-speed sequential gearbox has just one clutch for faster changes and a choice of seven different shift speeds. Dry sump lubrication, alloy sub-frames, and a rear transaxle highlight the LFA's F1 origins. With such cutting-edge technology, even at £343,000, Lexus lost money on every LFA they built.

LIMITED EDITION

A team of 175 engineers built the LFA in a dedicated factory, turning out one car a day. Numbers were limited to only 500 units and customers were specially chosen because they would not resell their cars for a profit. The last examples built were Nürburgring spec, good for 562 bhp and the most expensive Japanese road cars ever sold.

SOUND SYMPHONY
Engine note is piped into cabin by twin sound ducts.

EYES FRONT
Engine is front-mounted for perfect weight distribution.

CARBON CAPTURE
Body is special carbon reinforced polymer for extreme lightness.

SPECIFICATIONS

MODEL Lexus LFA (2009)
PRODUCTION 500
BODY STYLE Two-door, two-seater coupé.
CONSTRUCTION Carbon-fibre, aluminium.
ENGINE 4,805cc V10.
POWER OUTPUT 552 bhp.
TRANSMISSION Six-speed sequential.
SUSPENSION Double wishbone, multi-link.
BRAKES Four-wheel ceramic discs.
MAXIMUM SPEED 325 km/h (202 mph)
0–60 MPH (0–96 KM/H) 3.6 sec
0–100 MPH (0–161 KM/H) 7.6 sec
A.F.C. 7.2 km/l (17 mpg)

WILD CHILD

The design was hugely radical for traditional Lexus and meant
to endow the brand with badly needed sex appeal and glamour.
Using carbon-fibre saved 100 kg (220.5 lb) over an alloy body
and helped give the LFA a halo of technological modernity.

TOOLED UP
*Carbon tub is made
using one of only two
laser looms in the world.*

STOPPING POWER
*Rear Brembo brakes
have four pistons and
ceramic discs.*

WING FORCE
*Speed-sensitive rear wing
rises at over 80.5 km/h
(50 mph) to aid down force.*

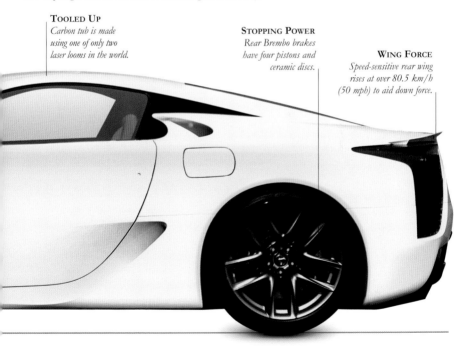

LINCOLN *Continental (1964)*

THERE'S AN UNSETTLING IRONY in the fact that John F. Kennedy was shot in a '61 Lincoln Continental. Like him, the revamped '61 Continental had a new integrity. Substantial and innovative, it was bristling with new ideas and survived for nine years without major change. The car fit for Presidents was elegant, restrained, and classically sculptured, perfect for Camelot's new dynasty of liberalism. Ironic, too, that JFK rather liked the Lincoln – he often used a stock White House Continental for unofficial business. Nearly $7,000 bought one of the most influential and best-built American cars of the Sixties. Not only did it carry a two-year, 39,000-km (24,000-mile) warranty, but every engine was bench-tested and each car given a 200-category shake-down. WASP America approved and production doubled in the first year. Even the Industrial Design Institute was impressed, awarding its coveted bronze medal for "an outstanding contribution of simplicity and design elegance".

LINEAR PROFILE

Apart from the gentle dip in the waistline at the back of the rear doors, the roof and wing lines form two uninterrupted, almost parallel lines. Low, wide, and mighty, the '60s Continental was considered the epitome of good taste and discrimination.

ENGINE

Power was supplied by a huge 430cid V8 that generated 320 bhp. Each engine was tested at near maximum revs for three hours and then stripped down for inspection.

INTERIOR

Every Continental had power steering and windows, walnut cappings, a padded dashboard, lush carpets, and vacuum-powered door locks as standard. The locks operated automatically as soon as the car started to move.

SPECIFICATIONS

MODEL Lincoln Continental Convertible (1964)

PRODUCTION 3,328

BODY STYLE Four-door, five-seater convertible.

CONSTRUCTION Steel body and chassis.

ENGINE 430cid V8.

POWER OUTPUT 320 bhp.

TRANSMISSION Three-speed Turbo-Drive automatic.

SUSPENSION *Front:* control arms and coil springs; *Rear:* leaf springs with live axle.

BRAKES Front and rear drums.

MAXIMUM SPEED 185 km/h (115 mph)

0–60 MPH (0–96 KM/H) 11 sec

A.F.C. 5 km/l (14 mpg)

SOLE RAG-TOP
When the revamped Conti was released in '61, Lincoln were the only manufacturer to offer a four-door convertible.

SHARED COSTS
To spread costs, the Continental shared some of its factory tooling with the '61 Thunderbird.

EASY ACCESS
The "suicide" rear-hinged doors hark back to classic pre-war coach-building. On older Continental Convertibles, opening all four doors at once can actually flex the floor and chassis.

CRUISE CONTROL
Even in '64 you could have cruise control, for a mere $96.

STEERING WHEEL
Least popular option in '64 was the adjustable steering wheel.

CONVERTIBLE RARITIES
Rag-top Continentals were really "convertible sedans" with standard power tops. The '64 rag-tops cost only $646 more than the four-door sedans, yet they remain much rarer: only about 10 per cent of all '61–'67 Lincolns produced were convertibles.

TYRES
Whitewalls were just one of numerous features that came as standard.

QUALITY NOT QUANTITY
The previous Conti was a leviathan, but not so the '61. The '61 restyle reflected the philosophy that big was not necessarily better.

SUSPENSION
Suspension damping was considered the best on any car.

STATE-OF-THE-ART HOOD
Eleven relays and a maze of linkages made the Continental's hood disappear neatly into the boot. The electrics were sealed and never needed maintenance. Along with the hood, the side glass and window frames also disappeared from view at the touch of a button.

CONSUMPTION
The Mark III Continental returned fuel figures of just 5 km/l (14 mpg).

NEW YORK
KTS 340

LOTUS *Elite*

IF EVER A CAR WAS A MARQUE landmark, this is it. The Elite was the first Lotus designed for road use rather than out-and-out racing, paving the way for a string of stunning sports and GT cars that, at the least, were always innovative. But the first Elite was much more than that. Its all-glass-fibre construction – chassis as well as body – was a bold departure that, coupled with many other innovations, marked the Elite out as truly exceptional, and all the more so considering the small-scale operation that created it. What's more, its built-in Lotus race-breeding gave it phenomenal handling and this, together with an unparalleled power-to-weight ratio, brought an almost unbroken run of racing successes. It also happens to be one of the prettiest cars of its era; in short, a superb GT in miniature.

CHAPMAN CREATION

The Elite was the brainchild of company founder and great racing innovator, Anthony Colin Bruce Chapman. The elegant coupé was a remarkable departure for the small company – and, to most, a complete surprise when it appeared at the London Motor Show in October 1957.

FILLER CAP
Quick-release fuel cap was an option many chose.

LOW DRAG
Low frontal area, with air intake below the bumper lip, helped Elite speed and economy. Drag coefficient was 0.29, a figure most other manufacturers would not match for 20 years.

HANDLE
Tiny door handle was little more than a hook.

WINDSCREEN
Concealed steel hoop around windscreen added stiffness and gave some roll-over protection.

RACE SUCCESS
Elites were uncatchable in their class, claiming Le Mans class wins six years in a row from 1959 to 1964.

WHEELS
48-spoke centre-lock Dunlop wire wheels were standard.

SPECIFICATIONS

MODEL Lotus Elite (1957–63)

PRODUCTION 988

BODY STYLE Two-door, two-seater sports coupé.

CONSTRUCTION Glass-fibre monocoque.

ENGINE Four-cylinder single ohc Coventry Climax, 1216cc.

POWER OUTPUT 75–105 bhp at 6100–6800 rpm.

TRANSMISSION Four-speed MG or ZF gearbox.

SUSPENSION Independent all round by wishbones and coil springs at front and MacPherson-type "Chapman strut" at rear.

BRAKES Discs all round (inboard at rear).

MAXIMUM SPEED 190 km/h (118 mph)

0–60 MPH (0–96 KM/H) 11.1 sec

A.F.C. 12.5 km/l (35 mpg)

AIR CHEATER

The Elite's aerodynamic make-up is remarkable considering there were no full-scale wind-tunnel tests, only low-speed air-flow experiments. The height of just 1.17 m (46 in) helped, as did the fully enclosed undertray below.

INTERIOR

Even tall owners were universal in their praise for driving comfort. The award winning interior was crisp and neat, with light, modern materials.

ECONOMY

Contemporary road tests recorded a remarkable 8.8 km/l (25 mpg) at a steady 161 km/h (100 mph).

SUSPENSION

Suspension was derived from the Lotus Formula 2 car of 1956.

LJC322

STRESSED ROOF
The roof was part of the Elite's stressed structure, which meant that popular calls for a convertible – especially from America – could not be answered. The solution came when the Elan was launched in 1962.

ROOF
SE (Special Equipment) models had silver roof as a "delete option".

BUMPERS
Both front and rear bumpers hid body moulding seams.

UNIT ORIGINS
Engine was developed from a wartime fire-pump engine.

ENGINE
The lightweight 1216cc four-cylinder engine was developed by Coventry Climax from their successful racing units. The unit's power rose from an initial 75 bhp to 83 bhp in the Elite's second series, but it was possible to extract over 100 bhp with options.

LOTUS *Elan Sprint*

THE LOTUS ELAN RANKS AS one of the best handling cars of its era. But not only was it among the most poised cars money could buy, it was also a thing of beauty. Conceived by engineering genius Colin Chapman to replace the race-bred Lotus 7, the Elan sat on a steel backbone chassis, clothed in a slippery glass-fibre body, and powered by a 1600cc Ford twin-cam engine. Despite a high price tag, critics and public raved and the Elan became one of the most charismatic sports cars of its decade, selling over 12,000 examples. Over an 11-year production life, with five different model series, it evolved into a very desirable and accelerative machine, culminating in the Elan Sprint, a 195 km/h (121 mph) banshee with a sub-seven second 0–60 (96 km/h) time. As one motoring magazine of the time remarked, "The Elan Sprint is one of the finest sports cars in the world". Praise indeed.

RACE ASSOCIATION
The duo-tone paintwork with dividing strip was a popular factory option for the Sprint. The red and gold combination had racing associations – the same livery as the Gold Leaf racing team cars. Everybody agreed that the diminutive Elan had an elfin charm.

INTERIOR
The Sprint's interior was refined and upmarket, with all-black trim, wood veneer dashboard, and even electric windows.

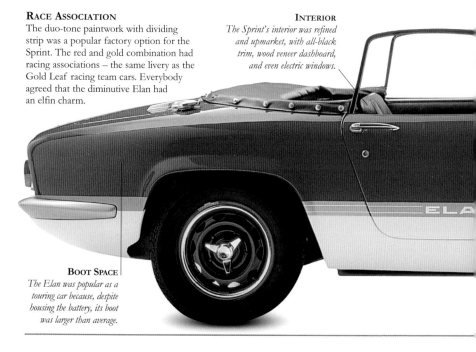

BOOT SPACE
The Elan was popular as a touring car because, despite housing the battery, its boot was larger than average.

SPECIFICATIONS

MODEL Lotus Elan Sprint (1970–73)

PRODUCTION 1,353

BODY STYLE Two-seater drophead.

CONSTRUCTION Steel box section backbone chassis.

ENGINE Four-cylinder twin overhead cam, 1558cc.

POWER OUTPUT 126 bhp at 6500 rpm.

TRANSMISSION Four-speed manual.

SUSPENSION Independent front and rear.

BRAKES Discs all round.

MAXIMUM SPEED 195 km/h (121 mph)

0–60 MPH (0–96 KM/H) 6.7 sec

0–100 MPH (0–161 KM/H) 15 sec

A.F.C. 8.5 km/l (24 mpg)

ENGINE

The "Big Valve" engine in the Sprint pushed out 126 bhp and blessed it with truly staggering performance. The Twin 40 DCOE Weber carburettors were hard to keep in tune.

STYLING

Perfectly proportioned from any angle, the Elan really looked like it meant business.

WRAP-AROUND BUMPERS

Front bumper was foam-filled glass-fibre and the Elan was one of the first cars to be fitted with bumpers that followed the car's contours.

BRAKES

Servo-assisted disc brakes provided tremendous stopping power.

MASERATI *Ghibli*

MANY RECKON THE GHIBLI is the greatest of all road-going Maseratis. It was the sensation of the 1966 Turin Show, and 30 years on is widely regarded as Maserati's ultimate front-engined road car – a supercar blend of luxury, performance, and stunning good looks that never again quite came together so sublimely on anything with the three-pointed trident. Pitched squarely against the Ferrari Daytona *(see page 233)* and Lamborghini Miura *(see page 318–21)*, it outsold both. Its engineering may have been dated, but it had the perfect pedigree, with plenty of power from its throaty V8 engine and a flawless Ghia design. It is an uncompromised supercar, yet it is also a consummate continent-eating grand tourer with 24-karat cachet. Muscular and perhaps even menacing, but not overbearingly macho, it is well mannered enough for the tastes of the mature super-rich. There will only be one dilemma; do you take the windy back roads or blast along the autoroutes? Why not a bit of both.

RACING STANCE

The Ghibli's dramatic styling is uncompromised, a sublime and extravagant 4.57 m (15 ft) of attitude that can only accommodate two people. From its blade-like front to its short, bobbed tail, it looks fast even in static pose. It has also aged all the better for its lack of finicky detail; the Ghibli's detail is simple and clean, worn modestly like fine, expensive jewellery.

WIDE VIEW

The front screen was huge but the mighty bonnet could make the Ghibli difficult to manoeuvre.

WHEELBASE

The Ghibli's wheelbase measured 255 cm (100 in).

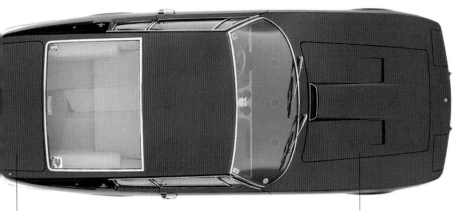

ACCELERATION
The mile (1.61 km) could be reached in just 15.1 seconds.

RAG-TOP CACHET
The most prized of all Ghiblis are the 125 convertible Spiders – out of a total Ghibli production figure of 1,274, only just over 100 were Spiders.

CARB-HEAVY
Four greedy twin-choke Weber carbs sat astride the V8.

THIRSTY
The Ghibli was a petrol gobbler, but when was there an economical supercar?

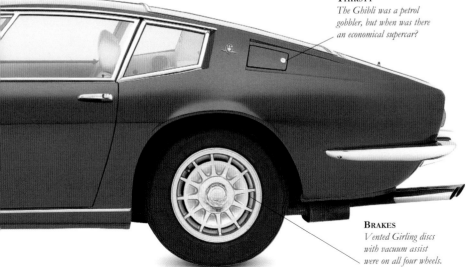

BRAKES
Vented Girling discs with vacuum assist were on all four wheels.

INTERIOR

A cliché certainly, but here you really feel you are on an aircraft flight-deck. The high centre console houses air-conditioning, which was standard Ghibli equipment. The steering wheel is adjustable and power steering was a later, desirable optional extra.

SPECIFICATIONS

MODEL Maserati Ghibli (1967–73)

PRODUCTION 1,274

BODY STYLES Two-door sports coupé or open Spider.

CONSTRUCTION Steel body and separate tubular chassis.

ENGINES Four-cam 90-degree V8, 4719cc or 4930cc (SS).

POWER OUTPUT 330 bhp at 5000 rpm (4719cc); 335 bhp at 5500 rpm (4930cc).

TRANSMISSION ZF five-speed manual or three-speed Borg-Warner auto.

SUSPENSION Wishbones and coil-springs at front; rigid axle with radius arms/semi-elliptic leaf springs at rear.

BRAKES Girling discs on all four wheels.

MAXIMUM SPEED 248 km/h (154 mph), 270 km/h (168 mph, SS)

0–60 MPH (0–96 KM/H) 6.6 sec, 6.2 sec (SS)

0–100 MPH (0–161 KM/H) 15.7 sec

A.F.C. 3.5 km/l (10 mpg)

UNDER THE BONNET

The potent race-bred quad-cam V8 is even-tempered and undemanding, delivering loads of low-down torque and accelerating meaningfully from as little as 500 rpm in fifth gear. This 1971 Ghibli SS has the 4.9-litre unit.

HEIGHT
At 118 cm (47 in), the Ghibli was a low sports coupé in the truest sense.

TRIDENT
Masers are instantly recognizable by the three-pointed trident.

EARLY GUIGIARO
Coachwork by Ghia was one of the finest early designs of their brilliant young Italian employee, Giorgetto Giugiaro. He was later to enhance his reputation with many other beautiful creations.

WHAT'S IN A NAME?
Like the earlier Mistral and the Bora, the Ghibli took its name from a regional wind. The Merak, which was introduced in 1972, was named after the smaller star of the constellation of the Plough. Other Maserati names were more race-inspired, including Indy, Sebring, and Mexico.

HIDE-AWAY HEADLIGHTS
Pop-up headlights might have improved looks when not needed, but they took their time to pop up. The Ghibli cost nearly $22,000 new in 1971 but buyers could be assured that they were getting a real deal supercar.

LIFT OFF
Wide front had a tendency to lift above 193 km/h (120 mph).

MASERATI *Kyalami*

THE 1970S PRODUCED some true automotive lemons. It was a decade when bare-faced badge engineering and gluttonous V8 engines were all the rage, and nobody cared that these big bruisers cost three arms and a leg to run. The Kyalami is one such monument to excess, a copy of the De Tomaso Longchamp with Maserati's all-alloy V8 on board instead of Ford's 5.8-litre cast-iron lump. The Kyalami was meant to compete with the Jaguar XJS but failed hopelessly. Plagued with electrical gremlins, this was a noisy, bulky, and unrefined machine that was neither beautiful nor poised. Yet for all that, it still sports that emotive trident on its nose and emits a deep and strident V8 bark. The Kyalami might not be a great car, but most of us, at least while looking at it, find it hard to tell the difference.

DE TOMASO ADAPTATION
Maserati designer Pietro Frua retouched the De Tomaso Longchamp design, turning it into the Kyalami. He gave it a new lower nose with twin lights, full width bonnet, and new rubber-cap bumpers with integral indicators.

REAR LIGHTS
Dainty rear light clusters were borrowed from the contemporary Fiat 130 Coupé.

NOT A PRETTY FACE

The frontal aspect is mean but clumsy. The three-part
front bumper looks cheap, while the Maserati grille
and trident seem to have been bolted
on as after-thoughts.

SPECIFICATIONS

MODEL Maserati Kyalami 4.9 (1976–82)

PRODUCTION 250 approx.

BODY STYLE Two-door, 2+2 sports saloon.

CONSTRUCTION Steel monocoque body.

ENGINE 4930cc all-alloy V8.

POWER OUTPUT 265 bhp at 6000 rpm.

TRANSMISSION Five-speed ZF manual
or three-speed Borg Warner automatic.

SUSPENSION Independent front with
coil springs and wishbones. Independent
rear with double coils, lower links, and
radius arms.

BRAKES Four-wheel discs.

MAXIMUM SPEED 237 km/h (147 mph)

0–60 MPH (0–96 KM/H) 7.6 sec

0–100 MPH (0–161 KM/H) 19.4 sec

A.F.C. 3.6 km/l (14 mpg)

STEERING
*Power-assisted steering robbed the car
of much needed accuracy and feel.*

ENGINE
*The engine was a four-cam, five-bearing
4.9 V8, with four twin-choke Weber
carbs, propelling the Kyalami to a
touch under 241 km/h (150 mph).*

TYRES
*The Kyalami generated
lots of commotion from
fat 205/70 Michelins.*

Mazda *RX7*

The RX7 arrived in American showrooms in 1978 and sales promptly went crazy. Even importing 4,000 a month, Mazda could not cope with demand and waiting lists were massive. For a while, RX7s changed hands on the black market for as much as $3,000 above retail price. By the time production ceased in 1985, nearly 500,000 had found grateful owners, making the RX7 the best-selling rotary car of all time. The RX7 sold on its clean European looks and Swiss-watch smoothness. Inspired by the woefully unreliable NSU Ro80 *(see pages 382–83)*, Mazda's engineers were not worried about the NSU's ghost haunting the RX7. By 1978 they had completely mastered rotary-engine technology and sold almost a million rotary-engined cars and trucks. These days the RX7 is becoming an emergent classic – the first car to make Felix Wankel's rotary design actually work and one of the more desirable and better made sports cars of the 1970s.

IMPRESSIVE AERODYNAMICS
The RX7's slippery, wind-cheating shape cleaved the air well, with a drag coefficient of only 0.36 and a top speed of 210 km/h (125 mph). Smooth aerodynamics helped the RX7 feel stable and composed with minimal body roll.

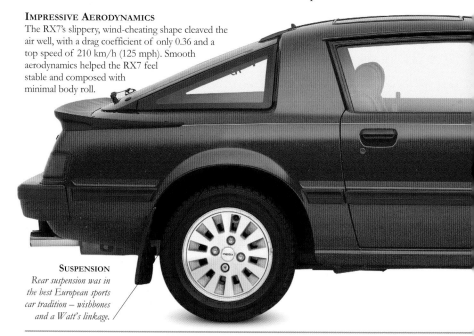

SUSPENSION
Rear suspension was in the best European sports car tradition – wishbones and a Watt's linkage.

SPOT-ON DESIGN

The body design was perfect from the start, and in its seven-year production run few changes were made to the slim and balanced shape.

SPECIFICATIONS

MODEL Maxda RX7 (1978–85)

PRODUCTION 474,565 (377,878 exported to US)

BODY STYLE All-steel coupé.

CONSTRUCTION One-piece monocoque bodyshell.

ENGINE Twin rotor, 1146cc.

POWER OUTPUT 135 bhp at 6000 rpm.

TRANSMISSION Five-speed all synchromesh/automatic option.

SUSPENSION Independent front. Live rear axle with trailing arms and Watt's linkage.

BRAKES *Front:* ventilated discs; *Rear:* drums.

MAXIMUM SPEED 210 km/h (125 mph)

0–60 MPH (0–96 KM/H) 8.9 sec

0–100 MPH (0–161 KM/H) 24 sec

A.F.C. 7.5 km/l (21.3 mpg)

HANDLING
Fine handling was due to near equal weight distribution and the low centre of gravity.

BONNET
The RX7's low bonnet line could not have been achieved with anything but the compact rotary engine, which weighed only 142 kg (312 lb).

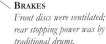

BRAKES
Front discs were ventilated; rear stopping power was by traditional drums.

FEDERAL IMPLICATIONS

The RX7 was originally planned as a two-seater, but Mazda was forced to include a small rear seat in the model. The reasoning behind this was that Japanese law stated all cars had to have more than two seats to encourage car sharing.

INTERIOR

Cockpit and dashboard are tastefully orthodox, with a handsome three-spoke wheel and five-gauge instrument binnacle. All UK-bound cars had five-speed manual transmission.

REAR PLANS

Original design plans for the RX7 favoured a one-piece rear tailgate like the Porsche 944, but economics dictated that an all-glass hatch was incorporated instead.

TURBO

The US could enjoy a brisk 217 km/h (135 mph) turbocharged model after 1984.

C418 DYV

ENGINE
The twin-rotor Wankel engine gave 135 bhp in later models. Reliable, compact, and easy to tune, there was even a small electric winch on the bulkhead to reel in the choke if owners forgot to push it back in.

ENGINE FLAWS
The Wankel-designed rotary engine had two weak points – low speed pull and fuel economy.

HEADLIGHTS
Pop-up headlights helped reduce wind resistance and add glamour. But, unlike those on the Lotus Esprit and Triumph TR7, the Mazda's always worked.

EUROPEAN STYLING
For a Japanese design, the RX7 was atypically European, with none of the garish over-adornment associated with other cars from Japan. Occasional rear seats and liftback rear window helped in the practicality department.

MERCEDES-BENZ *300SL Gullwing*

WITH ITS GORGEOUS GULLWING doors raised, the 300SL looked like it could fly. And with them lowered shut it really could, rocketing beyond 225 km/h (140 mph) and making its contemporary supercar pretenders look ordinary. Derived from the 1952 Le Mans-winning racer, these mighty Mercs were early forebears of modern supercars like the Jaguar XJ220 and McLaren F1 in taking race-track technology on to the streets. In fact, the 300SL can lay a plausible claim to being the first true post-war supercar. Awkward to enter, and with twitchy high-speed handling, it was sublimely impractical – it is a virtual supercar blueprint. It was a statement, too, that Mercedes had recovered from war-time devastation. Mercedes were back, and at the pinnacle of that three-pointed star was the fabulous 300SL, the company's first post-war sports car.

AERODYNAMICS
Detailed attention to aerodynamics was streets ahead of anything else at the time and helped make the 300SL the undisputed fastest road car of its era. Road cars developed 240 bhp, more than the racing versions of two years earlier.

WHEELS
Some say steel discs were fitted to keep costs down, but they also look more muscular than wires.

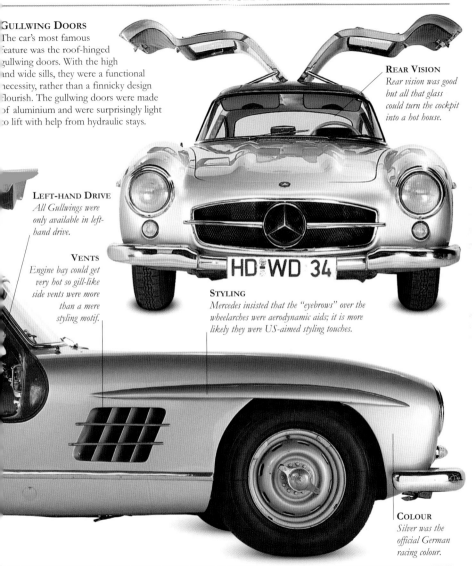

GULLWING DOORS

The car's most famous feature was the roof-hinged gullwing doors. With the high and wide sills, they were a functional necessity, rather than a finnicky design flourish. The gullwing doors were made of aluminium and were surprisingly light to lift with help from hydraulic stays.

REAR VISION

Rear vision was good but all that glass could turn the cockpit into a hot house.

LEFT-HAND DRIVE

All Gullwings were only available in left-hand drive.

VENTS

Engine bay could get very hot so gill-like side vents were more than a mere styling motif.

STYLING

Mercedes insisted that the "eyebrows" over the wheelarches were aerodynamic aids; it is more likely they were US-aimed styling touches.

COLOUR

Silver was the official German racing colour.

SPECIFICATIONS

Model Mercedes-Benz 300SL (1954–57)

Production 1,400

Body Style Two-door, two-seat coupé.

Construction Multitubular space-frame with steel and alloy body.

Engine Inline six-cylinder overhead camshaft, 2996cc.

Power Output 240 bhp at 6100 rpm.

Transmission Four-speed all synchromesh gearbox.

Suspension Coil springs all round, with double wishbones at front, swinging half-axles at rear.

Brakes Finned alloy drums.

Maximum Speed 217–265 km/h (135–165 mph), depending on gearing.

0–60 mph (0–96 km/h) 8.8 sec

0–100 mph (0–161 km/h) 21.0 sec

A.F.C. 6.4 km/l (18 mpg)

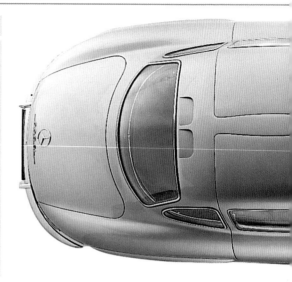

Smooth Rear

The Gullwing's smooth styling extended to the uncluttered rear; the boot lid suggests ample space, but this was not the case. The cockpit became quite hot, but air vents above the rear window helped.

Limited Space

As this sales illustration shows, with the spare tyre mounted atop the fuel tank there was very little room for luggage in the Gullwing's boot.

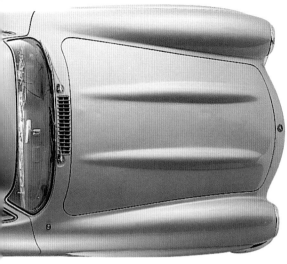

300SL ROADSTER
As Gullwing production wound down, Mercedes introduced the 300SL Roadster, which from 1957 to 1963 sold 1,858, compared to the Gullwing's 1,400. From 1955 to 1963 the 190SL Roadster served as the "poor man's" 300SL.

SLANT SIX
The engine was canted at 50 degrees to give a low bonnet-line. It was also the first application of fuel injection in a production car.

POWER SOURCE
The engine was originally derived from the 300-Series 3-litre saloons, then developed for the 1952 300SL racer, and two years later let loose in the road-going Gullwing, with fuel injection in place of carburettors.

BULGES
One bonnet bulge was for air intakes, the other for aesthetic balance.

STAR IDENTITY
The massive three-pointed star dominated the frontal aspect and was repeated in enamel on the bonnet edge.

TILT WHEEL
On some cars, mostly for the US, the wheel tilted to ease access.

HD·WD·34

MERCEDES-BENZ *280SL*

THE MERCEDES 280SL HAS mellowed magnificently. In 1963, the new SLs took over the sporting mantle of the ageing 190SL. They evolved from the original 230SL, through the 250SL, and on to the 280SL. The most remarkable thing is how modern they look, for with their uncluttered, clean-shaven good looks, it is hard to believe that the last one was made in 1971. Underneath the timelessly elegant sheet metal, they were based closely on the earlier Fintail saloons, sharing even the decidedly unsporting recirculating-ball steering. Yet it is the looks that mark this Merc out as something special, and the enduring design includes its distinctive so-called pagoda roof. This well-manicured Merc is a beautifully built boulevardier that will induce a sense of supreme self satisfaction on any journey.

SUSPENSION
Suspension was on the soft side for string-backed glove types.

TRADEMARK LIGHTS
So-called "stacked" headlights are unmistakeable Mercedes trademarks. Each outer lens concealed one headlamp, indicator, and sidelights.

FRENCH DESIGN
Design of the 280SL was down to Frenchman
Paul Bracq. Some macho types may dismiss it
as a woman's car and it is certainly not the most
hairy-chested of sporting Mercs.

GEARING
*Relatively few cars
were ordered with
a manual gearbox.*

OPTIONAL THIRD
*The SL was essentially
a two-seater, although a
third, sideways-facing rear
seat was available as a
(rare) optional extra.*

CLAP HANDS
*Windscreen wipers were
of the characteristic
"clap hands" pattern
beloved of Mercedes.*

HORN RING
*The D-shaped horn
ring allowed an
unobstructed view
of the instruments.*

CHROME BUMPER
The full-width front bumper
featured a central recess just
big enough for a standard
British number plate; the
quality of the chrome, as
elsewhere on the car,
was first class.

SL MOTIF
In Mercedes code-speak, the
S stood for Sport or Super,
L for Leicht (light) and
sometimes Luxus (luxury),
although at well over
1,362 kg (3,000 lb) it was
not particularly light.

SPECIFICATIONS

MODEL Mercedes-Benz 280SL (1968–71)

PRODUCTION 23,885

BODY STYLE Two-door, two-seat convertible with detachable hardtop.

CONSTRUCTION Pressed-steel monocoque.

ENGINE 2778cc inline six; two valves per cylinder; single overhead camshaft.

POWER OUTPUT 170 bhp at 5750 rpm.

TRANSMISSION Four- or five-speed manual, or optional four-speed auto.

SUSPENSION *Front:* independent, wishbones, coil springs, telescopic dampers; *Rear:* swing axle, coil springs, telescopic dampers.

BRAKES Front discs, rear drums.

MAXIMUM SPEED 195 km/h (121 mph, auto)

0–60 MPH (0–96 KM/H) 9.3 sec

0–100 MPH (0–161 KM/H) 30.6 sec

A.F.C. 6.7 km/l (19 mpg)

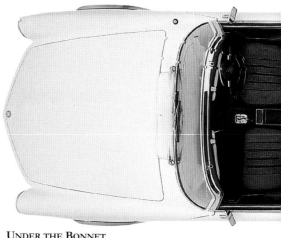

SAFE SUSPENSION
Swing-axle rear suspension was tamed to provide natural understeer.

UNDER THE BONNET
The six-cylinder ohc engine saw a process of steady development – the 2281cc 230SL in 1963, the 2496cc 250SL from 1966, and the final 2778cc 280SL shown here from 1968.

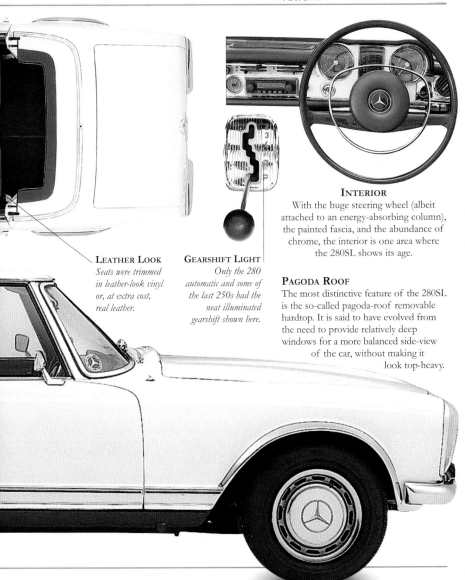

INTERIOR
With the huge steering wheel (albeit attached to an energy-absorbing column), the painted fascia, and the abundance of chrome, the interior is one area where the 280SL shows its age.

PAGODA ROOF
The most distinctive feature of the 280SL is the so-called pagoda-roof removable hardtop. It is said to have evolved from the need to provide relatively deep windows for a more balanced side-view of the car, without making it look top-heavy.

LEATHER LOOK
Seats were trimmed in leather-look vinyl or, at extra cost, real leather.

GEARSHIFT LIGHT
Only the 280 automatic and some of the last 250s had the neat illuminated gearshift shown here.

MERCEDES-BENZ *SLS AMG*

THE SLS COUPÉ REALLY IS ONE of the world's coolest rides, and the car I chose to drive every day. A blisteringly fast, wild hot-rod that doesn't rely on electronic gizmos, it looks wonderfully menacing, has the best soundtrack of any V8, and generates universal warm approval. Like the 1954 300SL Gullwing (*see pages 356–59*) that inspired it, the SLS is cramped, hard-riding, and edgy at the limit but intoxicatingly rapid. The best handling and most dramatic Mercedes ever, it is a worthy opponent to the Italian supercar set but looks infinitely more separate and distinctive. Climbing in and out can be a challenge but those crazy doors are the SLS's party piece. This is outrageous automotive mischief at its very finest.

HAND CRAFTED
The 6.3-litre AMG V8 is a masterpiece and each engine is hand assembled in-house by one man. Tuned to deliver 112 bhp more than the standard 6,300cc unit, 0–100 mph (0–161 km/h) is dispatched in an amazing 8 seconds. The Getrag seven-speed automatic has four switchable modes along with paddle-shift manual. Driven carefully, an SLS can return 9.8 km/l (23 mpg).

HERITAGE TOUCH
Twin chrome slashes over brake cooling vent echo original 300SL.

MANUAL DOORS
Doors do not have electric motors to save weight.

LIGHT FANTASTIC
Lightweight body is a mix of aluminium and plastic panels but is extremely rigid.

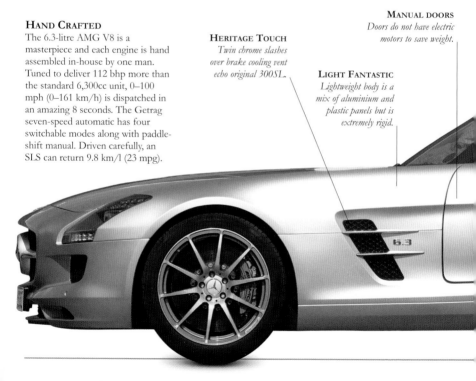

MIND YOUR HEAD
Gullwing doors are heavy and need long arms to close.

GO CAREFULLY
Wider than an S-Class Mercedes, Jaguar XJ, or Range Rover, the SLS can be hard work driving around town. Delicate AMG alloy wheels are vulnerable to damage from high kerbs and cost £2,500 ($3,720) a piece to replace.

SPECIFICATIONS

MODEL Mercedes-Benz SLS AMG (2010)

PRODUCTION N/A

BODY STYLE Two-seater coupé with gullwing doors.

CONSTRUCTION Alloy body panels with plastic boot lid.

ENGINE 6,208cc, V8.

POWER OUTPUT 563 bhp.

TRANSMISSION Seven-speed, dual-clutch semi-automatic.

SUSPENSION Double wishbone, coil spring.

BRAKES Four-wheel discs, optional ceramic.

MAXIMUM SPEED 317 km/h (197 mph)

0–60 MPH (0–96 KM/H) 3.8 sec

0–100 MPH (0–161 KM/H) 8 sec

A.F.C. 9 km/l (21 mpg)

STOPPING POWER
Orange caliper ceramic brakes are optional and help reduce fade on the track.

MG *TC Midget*

EVEN WHEN IT WAS NEW, the MG TC was not new. Introduced in September 1945, it displayed a direct lineage back to its pre-war forbears. If you were a little short on soul, you might even have called it old fashioned. Yet it was a trail-blazer, not in terms of performance, but in opening up new export markets. Popular myth has it that American GIs stationed in England cottoned on to these quaint sporting devices and when they got home were eager to take a little piece of England with them. Whatever the reality, it was the first in a long line of MG export successes. There was simply nothing remotely like this TC tiddler coming out of Detroit. It had a cramped cockpit, harsh ride, and lacked creature comforts, but when the road got twisty the TC could show you its tail and leave soft-sprung sofa-cars lumbering in its wake. It was challenging to drive, and all the more rewarding when you got it right.

EXHAUST
Rorty exhaust note was music to the ears.

TRADITIONAL CLASSIC

With its square-rigged layout, the TC is traditional with a capital T, and certainly a "classic" before the term was applied to cars. With its square front and separate headlamps, sweeping front wings, and cut-away doors, it is a true classic.

RAW MOTORING

While the TC may have been short on sophistication, it contained essential elements, such as wind-in-your-hair motoring, that marked it out as a true enthusiast's sporting car in the car-starved late 1940s.

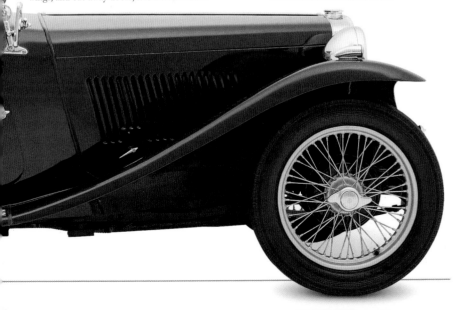

ENGINE
Ease of accessibility and maintenance was another of the TC's attractions. The XPAG engine was first used on some TB Midgets in 1939, then became standard MG wear until replaced by a 1500cc version in 1955. The TC was a popular race car, especially in the US, where it launched many careers and one world champion, Phil Hill.

OVERSEAS WINNER
Two TCs were exported for every one sold at home.

COCKPIT
Roomier than earlier Midgets, the TC cockpit was still cramped by comparison with less sporting contemporaries.

RIGHT-HOOKERS
Although over 2,000 were sold in America, all TCs were right-hand drive.

INTERIOR
Big Jaeger dials were in true British sporting tradition; the driver got the rev-counter, while speedo was in front of the passenger. A warning light on the dashboard – to the left of the speedo – illuminated if you exceeded Britain's 48 km/h (30 mph) urban speed limit.

SPECIFICATIONS

MODEL MG TC Midget (1947–49)

PRODUCTION 10,000

BODY STYLE Two-door, two-seater sports.

CONSTRUCTION Channel-section ladder-type chassis; ash-framed steel body.

ENGINE Four-cylinder overhead valve 1250cc, with twin SU carburettors.

POWER OUTPUT 54 bhp at 5200 rpm.

TRANSMISSION Four-speed gearbox with synchromesh on top three.

SUSPENSION Rigid front and rear axles on semi-elliptic springs, lever-type shock absorbers.

BRAKES Lockheed hydraulic drums.

MAXIMUM SPEED 117 km/h (73 mph)

0–60 MPH (0–96 KM/H) 22.7 sec

A.F.C. 9.9 km/l (28 mpg)

CONTINUED SUCCESS

The export trend begun so successfully by the TC really took off with the TD, which sold three times the number.

REPLACEMENT TD

The TC was replaced by the TD which, with its smaller disc wheels, chrome hub-caps, and bumpers, some MG aficionados considered less pure.

BRAKES

Lockheed drum brakes balanced the limited power output.

MGA

LAUNCHED IN SEPTEMBER 1955, the MGA was the first of the modern sporting MGs. The chassis, engine, and gearbox were all new, as was the smooth, Le Mans-inspired bodywork. Compared to its predecessor – the TF, which still sported old-fashioned running boards – the MGA was positively futuristic. Buyers thought so too, and being cheaper than its nearest rivals, the Triumph TR3 and Austin Healey 100, helped MG sell 13,000 cars in the first year of the MGA's production. The company's small factory at Abingdon, near Oxford, managed to export a staggering 81,000 MGAs to America. The car also earned an enviable reputation in competition, with the Twin Cam being the most powerful of the MGA engines.

VENTILATION
The chromed, shroud-panel vents at the front were for engine bay ventilation.

ENGINE
The tough B-Series, push-rod engine went well and lasted forever. A heater unit in front of the bulkhead was an optional extra. The 1600 model pushed out 80 bhp and featured front-disc brakes.

BS

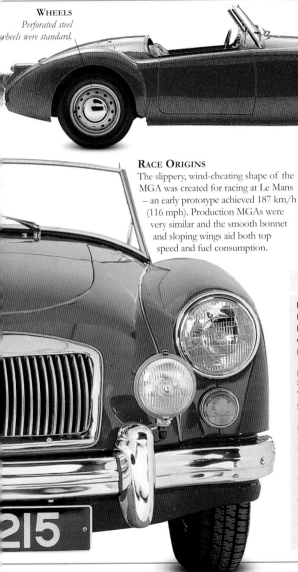

WHEELS
*Perforated steel
wheels were standard.*

WHEELS
*Perforated steel
wheels were standard.*

MATERIALS
*Door skins, bonnet, and
boot were light alloy.*

RACE ORIGINS

The slippery, wind-cheating shape of the MGA was created for racing at Le Mans – an early prototype achieved 187 km/h (116 mph). Production MGAs were very similar and the smooth bonnet and sloping wings aid both top speed and fuel consumption.

CONSTRUCTION

MGAs had a separate chassis, with the body bolted on top. The bodies were welded, painted, and trimmed at Morris Bodies in Coventry and then transported to Abingdon for the final fitting of mechanical equipment.

SPECIFICATIONS

MODEL MGA (1955–62)

PRODUCTION 101,081

BODY STYLE Two-door sports coupé.

CONSTRUCTION Steel.

ENGINES Four-cylinder 1489cc, 1588cc, 1622cc (Twin Cam).

POWER OUTPUT 72 bhp, 80 bhp, 85 bhp.

TRANSMISSION Four-speed manual.

SUSPENSION *Front:* independent; *Rear:* leaf-spring.

BRAKES Rear drums, front discs. All discs on De Luxe and Twin Cam.

MAXIMUM SPEED 161 km/h (100 mph); 181 km/h (113 mph) (Twin Cam).

0–60 MPH (0–96 KM/H) 15 sec (13.3 sec, Twin Cam)

0–100 MPH (0–161 KM/H) 47 sec (41 sec, Twin Cam)

A.F.C. 7–8.8 km/l (20–25 mpg)

MG*B*

WIDELY ADMIRED FOR ITS uncomplicated nature, timeless good looks, and brisk performance, the MGB caused a sensation back in 1962. The now famous advertising slogan "Your mother wouldn't like it" was quite wrong. She would have wholeheartedly approved of the MGB's reliability, practicality, and good sense. In 1965 came the even more practical tin-top MGB GT. These were the halcyon days of the MGB – chrome bumpers, leather seats, and wire wheels. In 1974, in pursuit of modernity and American safety regulations (the MGB's main market), the factory burdened the B with ungainly rubber bumpers, a higher ride height, and garish striped nylon seats, making the car slow, ugly, and unpredictable at the limit. Yet the B went on to become the best-selling single model sports car ever, finding 512,000 grateful owners throughout the world.

SIMPLE MECHANICS
All MGBs had the simple 1798cc B-series four-cylinder engine with origins going back to 1947. This Tourer's period charm is enhanced by the rare Iris Blue paintwork and seldom seen pressed-steel wheels – most examples were fitted with optional spoked wire wheels.

HOOD
Early cars had a "packaway" hood made from ICI Everflex.

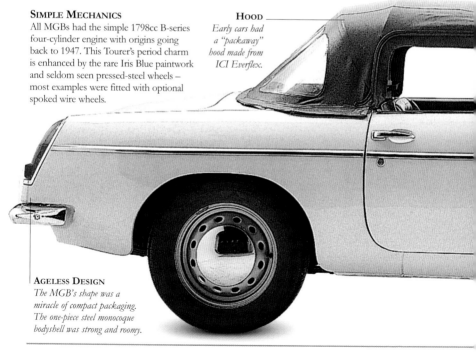

AGELESS DESIGN
The MGB's shape was a miracle of compact packaging. The one-piece steel monocoque bodyshell was strong and roomy.

SPECIFICATIONS

MODEL MGB Tourer (1962–1980)

PRODUCTION 512,243

BODY STYLE Steel front-engined two seater with aluminium bonnet.

CONSTRUCTION One-piece monocoque bodyshell.

ENGINE Four-cylinder 1798cc.

POWER OUTPUT 92 bhp at 5400 rpm.

TRANSMISSION Four-speed with overdrive.

SUSPENSION *Front:* independent coil; *Rear:* half-elliptic leaf springs.

BRAKES Lockheed discs front, drums rear.

MAXIMUM SPEED 171 km/h (106 mph)

0–60 MPH (0–96 KM/H) 12.2 sec

0–100 MPH (0–161 KM/H) 37 sec

A.F.C. 8.8 km/l (25 mpg)

INSIDE THE MG

The interior was vintage traditionalism at its best. Leather seats, crackle black metal fascia, nautical-sized steering wheel, and minor controls were strewn about the dash like boulders with scant thought for ergonomics.

MIRROR SUPPORT
The line down the centre of the windscreen was a mirror support rod.

BONNET
Bonnet was made out of lightweight aluminium.

SUSPENSION
Front suspension was coil spring with wishbones, and dated back to the MG TF of the 1950s.

MORGAN *Plus Four*

IT IS REMARKABLE THAT MORGANS are still made, but there is many a gent with a cloth cap and corduroys who is grateful that they are. Derived from the first four-wheeled Morgans of 1936, this is the car that buoyed Morgan on after the war while many of the old mainstays of the British motor industry wilted around it. Tweedier than a Scottish moor on the first day of the grouse shooting season, it is as quintessentially English as a car can be. It was a hit in America and other foreign parts, and it has also remained the backbone of the idiosyncratic Malvern-based company, which refuses to move with the times. Outdated and outmoded, Morgans are still so admired they hardly depreciate at all. First introduced in 1951, the Plus Four, with a series of Standard Vanguard and Triumph TR engines, laid the foundations for the modern miracle of the very old-fashioned Morgan Motor Company.

MODERN MORGAN
The second-generation Plus Four was the first of what are generally considered the "modern-looking" Morgans – if that is the right expression for a basic design which, still in production today, dates back to 1936.

"SUICIDE" DOORS
The earlier two-seat drophead coupé retained rear-hinged "suicide" doors; sports models had front-hinged doors.

ON THE RACK
Morgans have limited luggage capacity, so many owners fitted external racks.

REAR ILLUMINATION

Rear lights have never been a Morgan strong point. Amber indicators are a good 15 cm (6 in) inboard of the stop/tail lamps, and partially obscured by the luggage rack.

SPECIFICATIONS

MODEL Morgan Plus Four (1951–69)

PRODUCTION 3,737

BODY STYLES Two- and four-seater sports convertible.

CONSTRUCTION Steel chassis, ash frame, steel and alloy outer panels.

ENGINES 2088cc overhead-valve inline four (Vanguard); 1991cc or 2138cc overhead-valve inline four (TR).

POWER OUTPUT 105 bhp at 4700 rpm (2138cc TR engine).

TRANSMISSION Four-speed manual.

SUSPENSION *Front:* sliding stub axles, coil springs, and telescopic dampers; *Rear:* live axle, semi-elliptic leaf springs, and lever-arm dampers.

BRAKES Drums front and rear; front discs standard from 1960.

MAXIMUM SPEED 161 km/h (100 mph)

0–60 MPH (0–96 KM/H) 12 sec

A.F.C. 7–7.8 km/l (20–22 mpg)

SUSPENSION
The Plus Four retained simple sliding-pillar front suspension.

INTERIOR
From 1958, Plus Fours had a slightly wider cockpit with a new fascia. Speedometer, switches, warning lights, and minor gauges were grouped in a central panel on the dash.

HOOD
Unlike most convertible cars, the Plus Four has a hood which can be partially folded back.

REVISED FEATURES
Major distinguishing features on the second-generation Morgan include the cowled radiator grille and, from 1959, a wider body (as here) to provide more elbow room for driver and passenger. The doors were the only sensible places for external rear-view mirrors.

326 EPW

LIGHT WORK
Headlights are big, bold affairs set in pods on the front wings, but sidelights are about as visible as a pair of glow-worms.

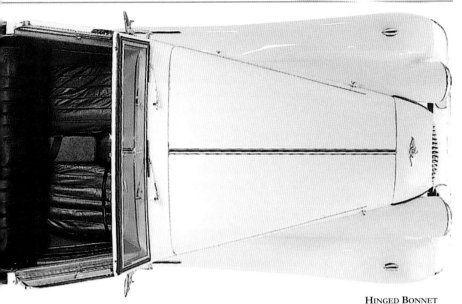

TRADITIONAL ASH FRAME

The current four-cylinder Morgan is built in exactly the same manner as most of its predecessors. The chassis is made from "Z"-section steel members, and on it sits a 94- or 114-piece wooden framework (two- and four-seat cars, respectively) clothed in a mixture of steel and aluminium panels. Today the company builds just two cars: the Plus Four and the Plus Eight.

ENGINE

The later Triumph TR3A 2138cc engine, as here, provided increased torque. The 2138cc engine was available in the TR3A from summer 1957. The earlier Triumph 1991cc engine was still available for those wishing to compete in sub-two-litre racing classes.

HINGED BONNET

Like all Morgans, the Plus Four has a hinged two-piece bonnet.

MORRIS *Minor MM Convertible*

THE MORRIS MINOR IS A motoring milestone. As Britain's first million seller it became a "people's car", staple transport for everyone from midwives to builders' merchants. Designed by Alec Issigonis, the genius who later went on to pen the Austin Mini *(see pages 44–47)*, the new Series MM Morris Minor of 1948 featured the then novel unitary chassis-body construction. The 918cc side-valve engine of the MM was rather more antique, a hang-over from the pre-war Morris 8. Its handling and ride comfort more than made up for the lack of power. With independent front suspension and crisp rack-and-pinion steering it embarrassed its rivals and even tempted the young Stirling Moss into high-speed cornering antics that lost him his licence for a month. Of all the 1.5 million Minors the most prized are the now rare Series MM convertibles.

RARE RAG-TOPS
Rag-tops remained part of the Minor model line-up until 1969, two years from the end of all Minor production. They represent only a small proportion of Minor production. Between 1963 and 1969 only 3,500 soft-tops were produced compared with 119,000 two-door saloons.

SUSPENSION
Morris bean counters dictated old-fashioned live-axle and leaf springs at the rear.

WINGS
Both front and rear wings were easily replaced, bolt-on items.

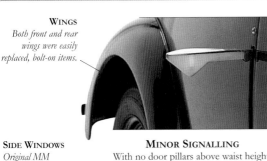

SPECIFICATIONS

MODEL Morris Minor (1948–71)

PRODUCTION 1,620,000

BODY STYLES Two- and four-door saloon, two-door convertible (Tourer), estate (Traveller), van, and pick-up.

CONSTRUCTION Unitary body/chassis; steel.

ENGINES Straight-four, 918cc, 803cc, 948cc, and 1098cc.

POWER OUTPUT 28 bhp (918cc); 48 bhp (1098cc).

TRANSMISSION Four-speed manual.

SUSPENSION Torsion bar independent front suspension; live-axle leaf-spring rear.

BRAKES Drums all round.

MAXIMUM SPEED 100–121 km/h (62–75 mph)

0–60 MPH (0–96 KM/H) 50+ sec for 918cc, 24 sec for 1098cc.

A.F.C. 12.7–15.2 km/l (36–43 mpg)

SIDE WINDOWS
Original MM Tourer had side curtains, replaced by glass rear windows in 1952.

MINOR SIGNALLING
With no door pillars above waist height, semaphore indicators were mounted lower down on the tourers; flashers eventually replaced semaphores in 1961. Few Minors today have their original semaphores.

ENGINE

The original 918cc side-valve engine was replaced progressively in 1952 and 1953 by the Austin A-series 803cc overhead valve engine, then by the A-series 948cc, and finally the 1098cc. Power outputs rose from 28 bhp on the 918 to 48 bhp on the 1098.

ENGINE ACCESS

Under-bonnet space and easy engine access make the Minor a DIY favourite.

"LOW LIGHTS"

In 1950, the headlights on all Minors were moved to the top of the wings. Earlier models such as the car featured here are now dubbed "low lights".

HANDLING

Even on cross-ply tyres the original Minor won praise for its handling; one journalist described it as "one of the fastest slow cars in existence".

LGO 786

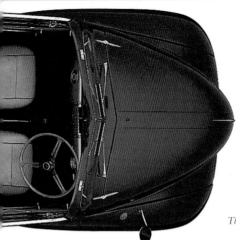

INTERIOR
This simple early
dashboard was
never really updated,
but the speedo was later
moved to the central console. The sprung-spoke
steering wheel was traditional, but rack-and-
pinion steering gave a crisp, light feel.

WINDSCREEN
*The split windscreen
was replaced by a
curved screen in 1956.*

ADVERTISING
*Sales literature described
the Minor as "The Best
Little Car in the World".*

FAKE CONVERTIBLES
So desirable are these open
tourers that in recent years
there has been a trade in
rogue rag-tops – chopped
saloons masquerading as
original factory convertibles.

MODEL CHOICE
At 155 cm (61 in) the
production car was 10 cm
(4 in) wider than the
prototype. At its launch the
Minor was available as a
two-door saloon and as a
convertible (Tourer). A
four-door, an estate, a
van, and a pick-up later
completed the range.

WIDENED BODY
*The fillet in the bumper
is another sign of the
widening of the body.*

NSU *Ro80*

ALONG WITH THE CITROËN DS *(see pages 178–81)*, the NSU Ro80 was 10 years ahead of itself. Beneath that striking, wind-cheating shape was an audacious twin-rotary engine, front-wheel drive, disc brakes, and a semi-automatic clutchless gearbox. In 1967, the Ro80 won the acclaimed "Car of the Year" award and went on to be hailed by many as "Car of the Decade". Technical pre-eminence aside, it also handled like a kart – the Ro80's stability, road-holding, ride, steering, and dynamic balance were exceptional, and far superior to most sports and GT cars. But NSU's brave new Wankel power unit was flawed and, due to acute rotor tip wear, would expire after only 24–32,000 km (15–20,000 miles). NSU honoured their warranty claims until they bled white and eventually Audi/VW took over, axing the Ro80 in 1977.

PASSENGER SPACE
With no transmission tunnel or propshaft, plenty of headroom, and a long wheelbase, rear passengers found the Ro80 thoroughly accommodating.

ROTARY RELIABILITY
Modern technology has made the troublesome Wankel engine reliable now, and prices of Ro80s have been creeping gently upwards.

FUTURISTIC DESIGN
In 1967, the Ro80 looked like a vision of the future with its low centre of gravity, huge glass area, and sleek aerodynamics. The high rear end, widely imitated a decade later, held a huge, deep boot.

MODEL NSU Ro80 (1967–77)

PRODUCTION 37,204

BODY STYLE Front-engine five-seater saloon.

CONSTRUCTION Integral chassis with pressed steel monocoque body.

ENGINE Two-rotor Wankel, 1990cc.

POWER OUTPUT 113.5 bhp at 5500 rpm.

TRANSMISSION Three-speed semi-automatic.

SUSPENSION Independent all round.

BRAKES Four-wheel discs.

MAXIMUM SPEED 180 km/h (112 mph)

0–60 MPH (0–96 KM/H) 11.9 sec

0–100 MPH (0–161 KM/H) 25 sec

A.F.C. 7 km/l (20 mpg)

UNDER THE BONNET
Designed by Felix Wankel, the brilliant twin-rotary engine was equivalent to a two-litre reciprocating piston unit. Drive was through a torque converter with a Fichel & Sachs electro-pneumatic servo to a three-speed NSU gearbox.

INTERIOR
Power steering was by ZF and the dashboard was a paragon of fuss-free Teutonic efficiency.

ENGINE POSITION
The engine was mounted on four progressive-acting rubbers with telescopic dampers on each side of the gearbox casing.

WHEELS
Stylish five-spoke alloys were optional equipment.

OLDSMOBILE *Toronado*

THE FIRST BIG FRONT-DRIVING land yacht since the Cord 810 of the Thirties, the Toronado was an automotive milestone and the most desirable Olds ever. With a 425cid V8 and unique chain-and-sprocket-drive automatic transmission, it had big-car power, outstanding road manners, and could crack 217 km/h (135 mph). Initial sales weren't brilliant, with sober buyers plumping for the more conventional Riviera, but by '71 the Riviera's design had lost its way and the Toronado really came into its own, selling up to 50,000 a year until the mid-Seventies. From then on, however, the more glamorous Cadillac Eldorado outsold both the Riviera and the Toronado. Built on an exclusive slow-moving assembly line, Toronados had few faults, which was remarkable for such a technically audacious car. Even so, the press carped about poor rear visibility and lousy gas mileage. But time heals all wounds, and these days there's no greater collector's car bargain than a '66–'67 Toronado.

DISTINCTIVE DESIGN
The Toro was a dream car design. Despite sharing a basic body with other GM models like the Riviera and Eldorado, it still emerged very separate and distinctive. *Automobile Quarterly* called it "logical, imaginative, and totally unique", and *Motor Trade* nominated it Car of the Year in 1966.

ENGINE HEAT
High engine temperatures and the huge Rochester 4GC four-barrel carb caused many under-bonnet fires.

PRICING
Standard sticker price was $4,585; de luxe versions ran to $4,779.

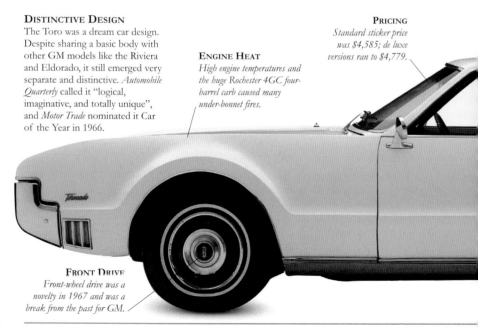

FRONT DRIVE
Front-wheel drive was a novelty in 1967 and was a break from the past for GM.

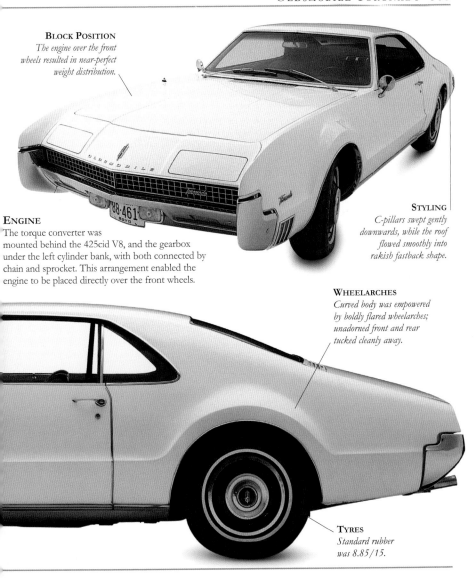

BLOCK POSITION
The engine over the front wheels resulted in near-perfect weight distribution.

ENGINE
The torque converter was mounted behind the 425cid V8, and the gearbox under the left cylinder bank, with both connected by chain and sprocket. This arrangement enabled the engine to be placed directly over the front wheels.

STYLING
C-pillars swept gently downwards, while the roof flowed smoothly into rakish fastback shape.

WHEELARCHES
Curved body was empowered by boldly flared wheelarches; unadorned front and rear tucked cleanly away.

TYRES
Standard rubber was 8.85/15.

TOP-FLIGHT CREDENTIALS
The Toronado was brisk, poised, and accurate. Understeer and front-wheel scrabble were kept to a minimum, and the car handled like a compact. Acceleration was in the Jaguar sedan league, and flat out it could chew the tail feathers of a Hi-Po Mustang.

REAR STYLING
Although an enormous car, the Toronado was a rakish fastback.

EXHAUSTS
Twin exhausts provided the outlet for the 425cid's grunt.

SPECIFICATIONS

MODEL Oldsmobile Toronado (1967)
PRODUCTION 21,790
BODY STYLE Two-door, five-seater coupé.
CONSTRUCTION Steel body and frame.
ENGINE 425cid V8.
POWER OUTPUT 385 bhp.
TRANSMISSION Three-speed Turbo Hydra-Matic automatic.
SUSPENSION *Front:* torsion bar; *Rear:* leaf springs with solid axle.
BRAKES Front and rear drums.
MAXIMUM SPEED 217 km/h (135 mph)
0–60 MPH (0–96 KM/H) 8.5 sec
A.F.C. 3.9 km/l (11 mpg)

INTERIOR
Standard equipment included Turbo Hydra-Matic tranny, power steering and brakes, Strato-bench front seat, de luxe armrests, rear cigarette lighters, foam seat cushions, and interiors in vinyl, leather, or cloth.

NOVEL FRONTAL STYLE

The concealed headlights and horizontal bar grille were genuinely innovative but would disappear in '68 for a heavier and less attractive front-end treatment. The Toronado's design arose in a free-expression competition organized by Olds in 1962. It became the marque's top model to date, and the equivalent of the Buick Riviera. The Toronado was GM's first commitment to front-wheel drive, which would become a corporate theology by 1980.

POP-UP LIGHTS

Unique retractable headlights were classic first-generation Toro.

OLDSMOBILE *4-4-2*

1971 WAS THE LAST OF THE 4-4-2's glory years. A performance package par excellence, it was GM's longest-lived muscle car, tracing its roots all the way back to the heady days of '64 when a 4-4-2 combo was made available for the Oldsmobile Cutlass F-85. Possibly some of the most refined slingshots ever to come from any GM division, 4-4-2s had looks, charisma, and brawn to spare. The 4-4-2 nomenclature stood for a four-barrel carb, four-speed manual transmission, and two exhausts. Olds cleverly raided the store room, using hot-shot parts previously only available to police departments. The deal was cheap and the noise on the street shattering. At $3,551, the super-swift Hardtop Coupe came with a 455cid V8, Rallye suspension, Strato bucket seats, and a top whack of 201 km/h (125 mph). The 4-4-2 package might have run and run had it not hit the '71 fuel crisis bang on. Which proved a shame, because it was to be a long time before power like this would be seen again.

PERFORMANCE ORIGINAL
From 1964 to '67, the 4-4-2 was simply a performance option that could be fitted into the F-85 range, but its growing popularity meant that in 1968 Olds decided to create a separate series for it in hardtop and convertible guises.

ENGINE BLOCK
Oldsmobile never tired of proclaiming that their 455cid mill was the largest V8 ever placed in a production car.

COLOUR CHOICES
In addition to this Viking Blue, Oldsmobile added Bittersweet, Lime Green, and Saturn Gold to their 1971 colour range.

MUSCLE LEGACY
Despite legislation that curbed the 4-4-2's power
output and led to the series being deleted after '71,
the 4-4-2 had made its mark and put Oldsmobile
well up there on the muscle-car map.

SPECIFICATIONS

MODEL Oldsmobile 4-4-2 (1971)

PRODUCTION 7,589 (1971)

BODY STYLES Two-door coupé and convertible.

CONSTRUCTION Steel body and chassis.

ENGINE 455cid V8.

POWER OUTPUT 340–350 bhp.

TRANSMISSION Three-speed manual, optional four-speed manual, three-speed Turbo Hydra-Matic automatic.

SUSPENSION *Front:* coil springs; *Rear:* leaf springs.

BRAKES Front discs, rear drums.

MAXIMUM SPEED 201 km/h (125 mph)

0–60 MPH (0–96 KM/H) 6.4 sec

A.F.C. 3.5–5 km/l (10–14 mpg)

REFLECTORS
Safety reflectors were evidence of an age where federal safety regulations were being introduced.

EXHAUST
Apart from the badge, the twin drain-pipe exhausts were the only clue that you were trailing a wild man.

POWER RESTRAINT
Unleaded fuel meant a drop in engine compression and therefore in speed.

INSIDE EXTRAS
The sports console at $77 and Rallye pack with clock and tacho at $84 were extras.

SALES PITCH
Advertising literature espoused the 4-4-2's torquey credentials: "A hot new number. Police needed it, Olds built it, pursuit proved it." The 4-4-2 was dropped completely from '81 to '84, but revived in '85, lasting until the final rear-wheel drive Cutlass was rolled out in '87.

MEDIA PRAISE
Motor Trend *said that "despite emission controls the '71 4-4-2 will still churn up plenty of smoke and fury".*

INTERIOR
Despite the cheap-looking, wood-grain vinyl dash, the 4-4-2's cabin had a real race-car feel. Bucket seats, custom steering wheel, and Hurst Competition gear shift came as standard.

MORE OPTIONS

1971 Cutlasses were offered in Convertible or Hardtop Coupe guise. 4-4-2s had bucket seats, wide-louvered bonnet, heavy-duty wheels, and super-wide bias-ply glass-belted tyres with white stripes. The hot $369 W-30 option included forced air induction, heavy-duty air cleaner, alloy intake manifold, body striping, sports mirrors, and special "W-car" emblems.

ENGINE

"Factory blue-printed to save you money", screamed the ads. The monster 455cid V8 was stock for 4-4-2s in '71, but it was its swansong year and power output would soon dwindle. By the late-Seventies, the 4-4-2 performance pack had been seriously emasculated.

OLDS FIGURES

In 1971 Olds churned out 558,889 cars, putting them in sixth place in the sales league.

REDUCED POWER

Sales literature pronounced that "4-4-2 performance is strictly top drawer", but in reality, unleaded fuel meant a performance penalty. Sixty could still be reached in under six seconds, though.

PACKARD *Hawk*

DISTINCTIVE, BIZARRE, AND VERY un-American, the '58 Hawk was a pastiche of European styling cues. Inspired by the likes of Ferrari and Mercedes, it boasted tan pleated-leather hide, white-on-black instruments, Jaguaresque wing vents, a turned metal dashboard, gulping bonnet air-scoop, and a broad glass-fibre shovel-nostril that could have been lifted off a Maserati. And it was supercharged. But Packard's attempt to distance themselves from traditional Detroit iron failed. At $4,000, the Hawk was overpriced, under-refined, and overdecorated. Packard had merged with Studebaker back in 1954, and although it was initially a successful alliance, problems with suppliers and another buy-out in 1956 basically sealed the company's fate. Only 588 Hawks were built, with the very last Packard rolling off the South Bend, Indiana, line on 13 July 1958. Today the Hawk stands as a quaint curiosity, a last-ditch attempt to preserve the Packard pedigree. It remains one of the most fiercely desired of the final Packards.

REAR ASPECT
Despite its European airs, no American car could escape the vogue for fins, and this car has two beauties. Nobody was too sure about the spare wheel impression on the boot, though.

ENGINE
Flight-O-Matic automatic transmission and a hefty, supercharged 289cid V8 came as standard. The Hawk's blower was a belt-driven McCulloch supercharger.

ATTRACTIVE PROFILE

Uniquely, the Hawk had exterior vinyl armrests running along the side windows and a refreshing lack of chrome gaudiness on the flanks. The roof line and halo roof band are aeronautical, and the belt line is tense.

SPECIFICATIONS

MODEL Packard Hawk (1958)

PRODUCTION 588 (1958)

BODY STYLE Two-door, four-seater coupé.

CONSTRUCTION Steel body and chassis.

ENGINE 289cid V8.

POWER OUTPUT 275 bhp.

TRANSMISSION Three-speed Flight-O-Matic automatic, optional overdrive.

SUSPENSION *Front:* independent coil springs; *Rear:* leaf springs.

BRAKES Front and rear drums.

MAXIMUM SPEED 201 km/h (125 mph)

0–60 MPH (0–96 KM/H) 8 sec

A.F.C. 5.3 km/l (15 mpg)

UNCONVENTIONAL FRONT

Even for the '50s, most buyers found the Hawk's frontal aspect a little too much, preferring instead the more traditional Detroit "million dollar chromium grin". The Hawk's styling was just plain ugly. And that's why it didn't sell.

PANHARD *PL17 Tigre*

PANHARD WAS ONE OF THE world's oldest names in car manufacturing, dating back to 1872. But by 1955 they had lost their upmarket image and had to be rescued by Citroën, who eventually bought them out completely in 1965. The Dyna, produced after the war in response to a need for a small, practical and economical machine, had an aluminium alloy frame, bulkhead, and horizontally opposed, air-cooled, twin-cylinder engine. In 1954, the Dyna became front-wheel drive, with a bulbous but streamlined new body. The 848cc flat-twin engine was a gem and in post-1961 Tigre guise pushed out 60 bhp; this gave 145 km/h (90 mph), enough to win a Monte Carlo Rally. Advertised as "the car that makes sense", the PL17 was light, quick, miserly on fuel, and years ahead of its time.

INTERIOR
The unusual interior had bizarre oval-shaped pedals, column change, and an unsuccessful pastiche of American styling themes.

STEERING
Technically advanced, the steering was rack-and-pinion, with only two turns lock-to-lock.

CYLINDER HEADS
Heads had hemispherical combustion chambers and valve-gearing incorporating torsion bars.

GALLIC AERODYNAMICS
With its aerodynamically shaped body, Panhard claimed the lowest drag coefficient of any production car in 1956. Emphasis was on weight-saving, with independent suspension and an aluminium frame and bulkhead. Despite its quirky Gallic looks, the PL17 was a triumph of outstanding efficiency.

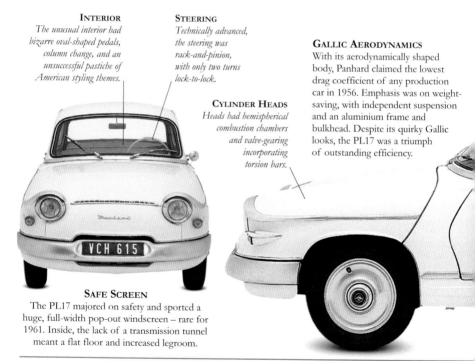

SAFE SCREEN
The PL17 majored on safety and sported a huge, full-width pop-out windscreen – rare for 1961. Inside, the lack of a transmission tunnel meant a flat floor and increased legroom.

ENGINE

The engine design dated back to 1940. Cylinders were cast integral with their heads in light alloy, cooling fins and cast-iron liners.

SPECIFICATIONS

MODEL Panhard PL17 Tigre (1961–64)

PRODUCTION 130,000 (all models)

BODY STYLE Four-door, four-seater sports saloon.

CONSTRUCTION Separate chassis with steel and aluminium body.

ENGINE 848cc twin horizontally-opposed air-cooled.

POWER OUTPUT 60 bhp at 5800 rpm.

TRANSMISSION Front-wheel drive four-speed manual.

SUSPENSION Independent front with twin transverse leaf, torsion bar rear.

BRAKES Four-wheel drums.

MAXIMUM SPEED 145 km/h (90 mph)

0–60 MPH (0–96 KM/H) 23.1 sec

A.F.C. 13.5 km/l (38 mpg)

EFFICIENT DESIGN

Simple design meant fewer moving parts, more power, and more miles to the gallon.

PEUGEOT *203*

COMPARED TO THE SCORES OF upright post-war saloons that looked like church pews, Peugeot's 203 was a breath of fresh air. As well as being one of the French car maker's most successful products, the 203's monocoque body and revolutionary engine set it apart. In its day, the 1290cc OHV power plant was state-of-the-art, with an aluminium cylinder head and hemispherical combustion chambers, said to be the inspiration for the famous Chrysler "Hemi" unit. With a range that included two- and four-door cabriolets, a family estate, and a two-door coupé, the French really took to the 203, loving its tough mechanicals, willing progress, and supple ride. By its demise in 1960, the 203 had broken records for Peugeot, with nearly 700,000 sold.

SHOW-STOPPER
Widely acclaimed at the 1948 Paris Motor Show, the 203's slippery shape was wind-tunnel tested in model form and claimed to have a rather optimistic drag coefficient of just 0.36 – lower than a modern Porsche 911 *(see pages 420–21)*. Quality touches abound, such as the exterior brightwork in stainless steel.

FUEL FILLER
This was concealed under a flush-fitting flap – unheard of in 1948.

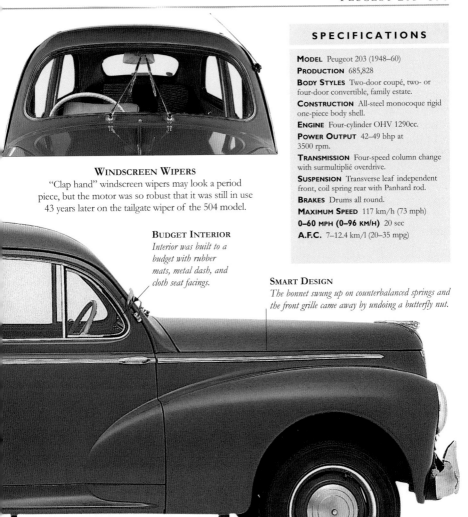

WINDSCREEN WIPERS
"Clap hand" windscreen wipers may look a period piece, but the motor was so robust that it was still in use 43 years later on the tailgate wiper of the 504 model.

BUDGET INTERIOR
Interior was built to a budget with rubber mats, metal dash, and cloth seat facings.

SMART DESIGN
The bonnet swung up on counterbalanced springs and the front grille came away by undoing a butterfly nut.

SPECIFICATIONS

MODEL Peugeot 203 (1948–60)

PRODUCTION 685,828

BODY STYLES Two-door coupé, two- or four-door convertible, family estate.

CONSTRUCTION All-steel monocoque rigid one-piece body shell.

ENGINE Four-cylinder OHV 1290cc.

POWER OUTPUT 42–49 bhp at 3500 rpm.

TRANSMISSION Four-speed column change with surmultiplié overdrive.

SUSPENSION Transverse leaf independent front, coil spring rear with Panhard rod.

BRAKES Drums all round.

MAXIMUM SPEED 117 km/h (73 mph)

0–60 MPH (0–96 KM/H) 20 sec

A.F.C. 7–12.4 km/l (20–35 mpg)

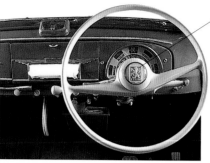

BADGE
Peugeot's lion badge dates back to 1906, when Robert Peugeot started up his own company called Lion-Peugeot.

INTERIOR
With post-war steel in short supply, aluminium was used to good effect in the under-dash handbrake and column gear change. The handsome fastback body gave plenty of cabin room.

ENDURING BLOCK
The basic design was still used in the 1980s for Peugeot's 1971cc 505 model.

ENGINE
The 49 bhp OHV push-rod engine was the 203's most advanced feature. With wet liners, low compression ratio, and alloy head, it was smooth, free-revving and long-lasting.

GEARBOX
The four-speed gearbox was really a three with overdrive.

RACK MOUNTS
Integral mounting points for a roof rack were a nice styling touch.

STYLISH BUTT
These stylish sweeping curves were influenced by the 1946 Chevrolet. A vast boot with a low-loading sill made the 203 ideal family transport. Another side to the 203 was racing; many were tuned and campaigned by privateers in rallies like the Monte Carlo.

PAINTWORK
A high gloss finish was achieved by the application of several coats of synthetic lacquer.

FRONT VIEW
The 203 was modified in 1953 with a curved windscreen, revised dashboard, and front quarter lights. This model was registered in 1955. The 203's turning circle was usefully tight – only 5.39 m (14 ft 9 in), with three turns lock-to-lock. Despite its 18 cwt weight and relatively modest power output, the handsome Peugeot's performance was sprightly.

SUSPENSION
Front suspension was by transverse leaf independent springing.

PLYMOUTH *Barracuda* (1964)

THE BIG THREE WEREN'T slow to cash in on the Sixties' youth boom. Ford couldn't keep their Mustang project secret and the Chrysler Corporation desperately wanted a piece of the action. But they had to work fast. They took their existing compact, the Plymouth Valiant, prettied up the front end, added a dramatic wrap-around rear window, and called it the Barracuda. It hit the showroom carpets in April 1964, a fortnight before the Mustang. A disarming amalgam of performance, poise, and refinement, Plymouth had achieved a miracle on the scale of loaves and fishes – they made the Barracuda fast, yet handle crisply and ride smoothly. The 273cid V8 made the car quicker than a Mustang, but that bizarre rear window dated fiercely and Mustangs outsold Barracudas 10-to-one. Plymouth believed the long-bonnet-short-boot "pony" formula wouldn't captivate consumers like a swooping, sporty fastback. Half a million Mustang buyers told them they'd backed the wrong horse.

HOT INSIDE
The greenhouse interior got hot on sunny days but was well detailed and enormously practical. Standard fare was bucket seats and bucket-shaped rear bench seat. Instruments were matt silver with circular chrome bezels. The padded dash was a $16.35 extra, as was a wood-grain steering wheel.

TRANSMISSION
Optional was Chrysler's new Hurst-linkage manual transmission.

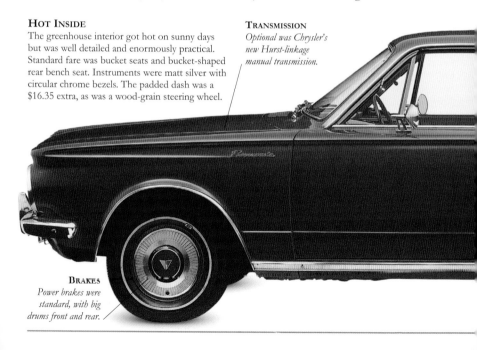

BRAKES
Power brakes were standard, with big drums front and rear.

REAR WINDOW
*Massive window
earned the 'Cuda
top marks for safety.*

ACRES OF GLASS
The fastback glass wrapped down to the rear wing line
and was developed by the Pittsburgh Plate Glass
Company; it was the largest use of glass in any
production car to date. As a result, visibility was epic.

SPECIFICATIONS

MODEL Plymouth Barracuda (1964)
PRODUCTION 23,443 (1964)
BODY STYLE Two-door fastback.
CONSTRUCTION Steel body and chassis.
ENGINES 170cid, 225cid sixes,
273cid V8.
POWER OUTPUT 101–235 bhp.
TRANSMISSION Three-speed manual,
optional four-speed manual, and
three-speed TorqueFlite automatic.
SUSPENSION *Front:* torsion bar;
Rear: leaf springs.
BRAKES Front and rear drums, optional
front discs.
MAXIMUM SPEED 161–177 km/h
(100–110 mph)
0–60 MPH (0–96 KM/H) 8–13 sec
A.F.C. 5.7–7.8 km/l (16–22 mpg)

RAG-TOP OPTION
*In '67 a convertible was
added with power hood
and real glass window.*

DIFFERENTIAL
*New Sure-Grip
differential was offered
as an extra to buyers.*

MUSTANG CONTRAST
Compared with the Mustang, the Barracuda's front was busy, cluttered, and lacked symmetry, but it was a brave and bold design. Had the Mustang not been launched in the same month, things might have been very different.

MIRROR
Remote-controlled outside wing mirror was a $12 convenience option.

ADJUSTABLE MIRROR
Prismatic day-and-night mirror could be adjusted to deflect annoying headlight glare at night.

SELLING THE WHEEL
The 'Cuda brochure insisted that the optional wood-grain steering wheel "gave you the feel of a racing car".

PLYMOUTH

19 400TH ANNIVERSARY 65
1W166142
FLORIDA

MEDIA PRAISE
Road and Track *magazine said, "for sports car performance and practicality, the Barracuda is perfect".*

VALIANT LINKS
The Barracuda was a Plymouth Valiant from the roof line down and shared its power and suspension.

FLEXIBLE SEAT
*Bucket seat could
be adjusted into
six positions.*

COLOURS
*Interior colours
available were gold,
blue, black, or this
smart red.*

BOOT SPACE
The rear seats folded forward to
produce an astronomical cargo
area that measured 2.14 m (7 ft)
long. Based on the mass-market,
best-selling Valiant, the Barracuda
was aimed at a completely new
market – rich young things with
a desire to look cool.

BUMPERS
*Bumper guards were
an $11.45 option.*

THE FORMULA S OPTION
The 'Cuda's base engine was a
170cid slant six. Other mills were
the 225cid six and two-barrel
273cid V8. Despite the fact that
the Formula S offered a V8
block plus race trimmings,
this was still rather tame by
Plymouth standards. The '61
Fury, for example, had a 318cid
unit that pushed out 230 bhp.

Did you know
that the 1965 Plymouth Barracuda
has an optional Formula 'S' sports package
that includes a Commando 273-cu.in.
V-8 engine; heavy-duty shocks, springs, and
sway bar; a tachometer; wide-rim (14-in.)
wheels, special Blue Streak tires, and
simulated bolt-on wheel covers?
You do now.

PLYMOUTH *'Cuda (1970)*

THE TOUGH-SOUNDING '70s 'Cuda was one of the last flowerings of America's performance binge. Furiously fast, it was a totally new incarnation of the first '64 Barracuda and unashamedly aimed at psychopathic street-racers. Cynically, Plymouth even dubbed their belligerent model line-up "The Rapid Transit System". '70 Barracudas came in three styles – the 'Cuda was the performance model – and nine engine choices, topped by the outrageous 426cid Hemi. Chrysler's advertising men bellowed that the Hemi was "our angriest body wrapped around ol' King Kong hisself". But rising insurance rates and new emission standards meant that the muscle car was an endangered species. By 1973 Plymouth brochures showed a 'Cuda with a young married couple, complete with a baby in the smiling woman's arms. The party was well and truly over.

NEAT DESIGN

The '70 'Cuda's crisp, taut styling is shared with the Dodge Challenger, and the classic long-bonnet-short-boot design leaves you in no doubt that this is a pony car. Government legislation and hefty insurance rates ensured that this was the penultimate year of the big-engined Barracudas; after '71, the biggest block on offer was a 340cid V8.

AIR CLEANER

Unsilenced air cleaners such as this weren't allowed in California because of drive-by noise regulations.

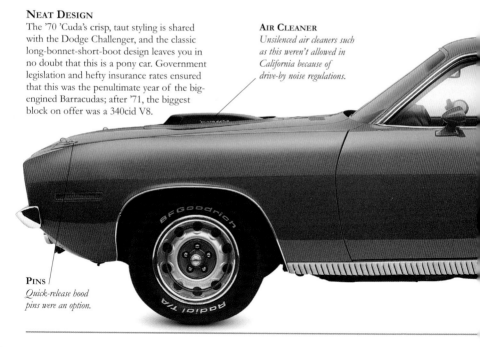

PINS
Quick-release hood pins were an option.

AIR CLEANER
The air cleaner vibrated ("shaked") through the top of the bonnet, a standard 'Cuda feature.

SALES FIGURES
Total 1970 'Cuda production was a healthy 30,267 units.

ENGINE
The 440cid "six-pack" Magnum motor cranked out 385 bhp and drank through three two-barrel Holley carbs, explaining the six-pack label. Base engine was a 383cid V8, which pushed out 335 horses.

SPECIFICATIONS

MODEL Plymouth 'Cuda (1970)

PRODUCTION 30,267 (1970)

BODY STYLES Two-door, four-seater coupé and convertible.

CONSTRUCTION Steel unitary body.

ENGINES 383cid, 426cid, 440cid V8s.

POWER OUTPUT 335–425 bhp.

TRANSMISSION Three-speed manual, optional four-speed manual, and three-speed TorqueFlite automatic.

SUSPENSION *Front:* torsion bars; *Rear:* leaf springs with live axle.

BRAKES Front discs, rear drums.

MAXIMUM SPEED 220–241 km/h (137–150 mph)

0–60 MPH (0–96 KM/H) 5.9–6.9 sec

A.F.C. 4.2–6 km/l (12–17 mpg)

PERFORMANCE PARTS
Super Stock springs and a heavy-duty Dana 60 rear axle were standard on all 440 'Cudas.

STRIPING
Optional inverted hockey stick graphics trumpeted engine size.

OVERHEAD STYLING

Plymouth stylists kept the shape
uncluttered, with tapered-in
bumpers, concealed wipers,
flush door handles, smooth
overhangs, and subtly flared
wheelarches. Even so, the 'Cuda
had ballooned in proportions
since the first Barracuda models
of the mid-Sixties and, along
with the Mustang *(see pages
278–85)*, now started to lose its
raison d'être. With the energy
crisis just around the corner,
its days were numbered.

HIDDEN WIPERS
*Windscreen wipers
were neatly concealed
behind the rear lip
of the bonnet.*

RACING MIRRORS
*Colour-coded racing
mirrors could be
ordered for $26.*

BIG-BLOCK SPEED
The 440-6 was a $250 'Cuda
engine option that allowed the
car to hit the quarter mile in
14.44 seconds. Only 652
1970 'Cuda hardtops were
fitted with the awesome
$871 Street Hemi V8.

TRANSMISSION
*Quick manual upshifts
were possible with the
Slap Stik T-handle.*

INTERIOR
'Cuda interiors were flamboyant,
with body-hugging bucket seats,
Hurst pistol-grip shifter, and wood-
grain steering wheel. This model has
the Rallye instrument cluster, with
tachometer and oil pressure gauge.

'CUDA BADGE
*'Cuda was a slang name
coined by Woodward
Avenue cruisers.*

COLOUR CHOICE
*'Cudas came in 18 strident colours,
with funky names like In Violet,
Lemon Twist, and Vitamin C.*

DECLINING FIGURES
Though 'Cuda hardtop
models cost $3,164 in 1970,
by '74, total Barracuda
sales for the year had
slipped to just over
11,000, and it was axed
before the '75 model year.

TWIN EXHAUSTS
*Provocative square
exhausts left no doubt
about the 'Cuda's grunt.*

PONTIAC *GTO*

"THE GREAT ONE" WAS Pontiac's answer to a youth market with attitude and disposable cash. Detroit exploited a generation's rebellion by creating cars with machismo to burn. In 1964, John DeLorean, Pontiac's chief engineer, shoe-horned the division's biggest V8 into the timid little Tempest compact with electrifying results. He then beefed up the brakes and suspension, threw in three two-barrel carbs, and garnished the result with a name that belonged to a Ferrari. In 1966 it became a model in its own right, and Detroit's first "muscle car" had been born. Pundits reckon the flowing lines of these second-generation GTOs make them the best-looking of all. Engines were energetic performers too, with a standard 335 bhp 389cid V8 that could be specified in 360 bhp high-output tune. But by '67 GTO sales had tailed off by 15 per cent, depressed by a burgeoning social conscience and federal meddling. The performance era was about to be legislated into the history books.

ORIGINAL MUSCLE

John DeLorean's idea of placing a high-spec engine in the standard Tempest body paved the way for a whole new genre and gave Pontiac immediate success in '64. Had Ford not chosen to release the Mustang in the same year, the GTO would have been the star of '64, and even more sales would have been secured.

BIG BLOCK

Pontiac were the first mainstream manufacturer to combine big-cube power with a light body. In tests, a '66 Convertible hit 60 mph (96 km/h) in 6.8 seconds.

WHEELS
Five-spoke Rally II sport wheels were a $72 option.

SALES SUCCESS
Sales peaked in 1966, with over 95,000 GTOs
going to power-hungry young drivers whose
average age was 25. The convertible was the
most aesthetically pleasing of the range.

SPECIFICATIONS

MODEL Pontiac GTO Convertible (1966)

PRODUCTION 96,946 (1966, all body styles)

BODY STYLES Two-door, five-seater
hardtop, coupé, and convertible.

CONSTRUCTION Steel unitary body.

ENGINE 389cid V8s.

POWER OUTPUT 335–360 bhp.

TRANSMISSION Three-speed manual,
optional four-speed manual, and three-speed
Hydra-Matic automatic.

SUSPENSION Front and rear coil
springs.

BRAKES Front and rear drums, optional
discs.

MAXIMUM SPEED 201 km/h
(125 mph)

0–60 MPH (0–96 KM/H) 6.8–9.5 sec

A.F.C. 5.3 km/l (15 mpg)

GTO IMAGE
*The GTO had a mischievous
image and was described as a
"methodist minister leaving
a massage parlour".*

LENGTH
*It might look long, but the
GTO was actually
38 cm (15 in) shorter than
Pontiac's largest models.*

PERFORMANCE REAR
*The GTO came with heavy-duty
shocks and springs as standard,
along with a stabilizer bar.*

CHOICE EXTRAS
*GTOs could be ordered
with Rally Cluster gauges,
close-ratio four-on-the-floor,
centre console, and walnut
grain dash insert.*

SEATS
*Reclining front seats
could be specified
as an extra.*

INTERIOR
GTOs were equipped to the same
high standard as the Pontiac
Tempest Le Mans. Items included
ashtray lights, cigarette lighter,
carpeting, and a power top
for convertibles. Air-
conditioning and power
steering could be
ordered at $343 and
$95 respectively.

HEADLIGHTS
*The stacked headlights
were new for Pontiacs in
'65 and were retained
on GTOs until the
end of the decade.*

NICKNAME
*Muscle-car buffs
dubbed the GTO
"The Goat".*

ENGINE OPTION
*The HO model could
do the standing quarter
in 14.2 seconds.*

GTO BADGE
Road & Track *magazine
reckoned the theft of the GTO
name from Ferrari was "an act
of unforgivable dishonesty".*

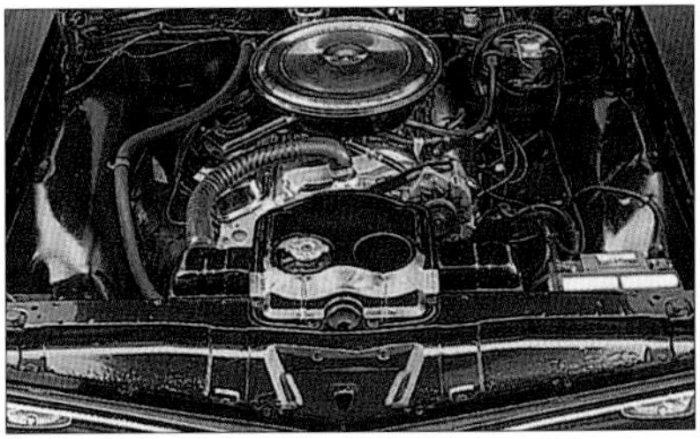

ENGINE
The base 335 bhp 389cid block had a
high-output Tri-Power big brother that
pushed out 360 bhp for an extra $116.
The range was expanded in '67 to
include an economy 255 bhp 400cid
V8 and a Ram-Air 400cid mill that
also developed 360 bhp, but at
higher revs per minute.

INDICATORS
*Turn signals in grille
were meant to mimic
European-style
driving lights.*

'66 FACELIFT
First-generation
GTOs were facelifted
in '66 with a more
aggressive split grille and
stacked headlight treatment
and gently kicked-up rear
wings. 1966 GTOs such
as the example here were
Pontiac's most popular,
with sales nudging close
to 100,000 units.

G T O

PONTIAC *Trans Am*

IN THE SEVENTIES, FOR THE FIRST TIME in American history, the Government intervened in the motor industry. With the 1973 oil crisis, the Big Three were ordered to tighten their belts. Automotive design came to a halt, and the big-block Trans Am became the last of the really fast cars. The muscular Firebird had been around since 1969 and, with its rounded bulges, looked as if its skin had been forced out by the strength underneath. Gas shortage or not, the public liked the '73 Trans Am, and sales quadrupled. The 455 Super Duty V8 put out 310 horsepower and, while Pontiac bravely tried to ignore the killjoy legislation, someone remarked that their High Output 455 was the largest engine ever offered in a pony car. The game was up, and within months modifications to comply with emission regulations had brought power down to 290 bhp. The hell-raising 455 soldiered on until 1976, and that athletic fastback body until '82. But the frenetic muscle years of 1967–73 had irretrievably passed, and those wonderful big-block banshees would never be seen again.

ESTABLISHED MUSCLE

Detroit's oldest warrior, the Firebird is the only muscle car that's been in the brochures for 30 years. Based on the Camaro's F-body, the Firebird debuted in 1967, but the wild Trans Am didn't appear until '69. Surprisingly, there was little fanfare until the hot 1970 restyle.

BONNET SCOOP

The rear-facing "shaker" bonnet scoop was an indication of the Trans Am's immense power.

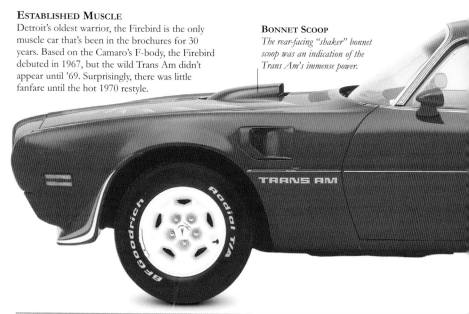

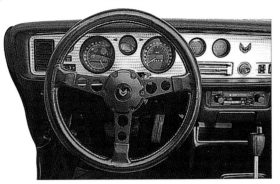

DASHBOARD

Second-edition Trans Ams had a standard engine-turned dash insert, Rally gauges, bucket seats, and a Formula steering wheel. The tacho was calibrated to a very optimistic 8000 rpm. The speedo was just as untruthful, with a maximum of 160 mph (257 km/h).

SPECIFICATIONS

MODEL Pontiac Firebird Trans Am (1973)

PRODUCTION 4,802 (1973)

BODY STYLE Two-door, four-seater fastback.

CONSTRUCTION Steel unitary body.

ENGINE 455cid V8.

POWER OUTPUT 250–310 bhp.

TRANSMISSION Four-speed manual or three-speed Turbo Hydra-Matic automatic.

SUSPENSION *Front:* coil springs; *Rear:* leaf springs with live axle.

BRAKES Front discs, rear drums.

MAXIMUM SPEED 217 km/h (135 mph)

0–60 MPH (0–96 KM/H) 5.4 sec

A.F.C. 6 km/l (17 mpg)

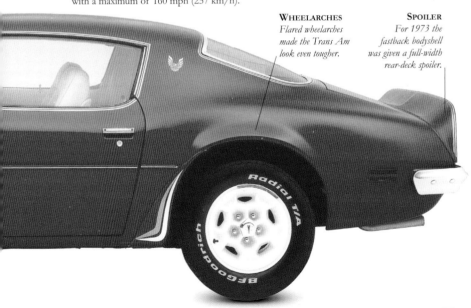

WHEELARCHES
Flared wheelarches made the Trans Am look even tougher.

SPOILER
For 1973 the fastback bodyshell was given a full-width rear-deck spoiler.

DECORATIVE DECAL

The "screaming chicken" graphics gracing the bonnet were new for 1973. Created by stylist John Schinella, they were a modern rendition of the Native American phoenix symbol. The Trans Am now looked as distinctive as it drove.

BODY BY FISHER

Pontiac wanted to portray that bodies were hand-built by an old-time carriage-maker.

'73 REVIVAL

Steep insurance rates and a national shift away from performance iron didn't help Trans Am sales, but in 1973, the year of the "screaming chicken" bonnet decal and Super Duty V8, Trans Ams left showrooms like heat-seeking missiles. Nearly killed off by GM, it soldiered on into the emasculated '80s and '90s.

FRONT VALANCE

New front valance panel with small air dam appeared in 1973.

ENGINE
The big-block Trans Ams were
Detroit's final salute to performance.
The 455 Super Duty could reach 60
(96 km/h) in under six seconds, and
run to 217 km/h (135 mph).

NAME IN DISPUTE
The Trans Am name was "borrowed"
from the Sports Car Club of
America, and the SCCA
threatened to sue unless
Pontiac paid a royalty of $5
per car. The Trans Am was a
seriously macho machine,
with *Car & Driver*
magazine calling it "a
hard-muscled, lightning-
reflexed commando
of a car".

EXHAUSTS
*Dual exhausts with
chrome extensions
were standard.*

PORSCHE *356B*

VW BEETLE DESIGNER Ferdinand Porsche may have given the world the "people's car", but it was his son Ferry who, with Karl Rabe, created the 356. These days a Porsche stands for precision, performance, purity, and perfection, and the 356 is the first chapter in that story. Well not quite. The 356 was so-named because it was the 356th project from the Porsche design office. It was also the first car to bear the Porsche name. Post-war expediency forced a reliance on Beetle underpinnings, but the 356 is much more than a Bug in butterfly's clothes. Its rear-engined layout and design descends from the father car, but in the athletic son the genes are mutated into a true sporting machine. A pert, nimble, tail-happy treat, the pretty 356 is the foundation stone of a proud sporting tradition.

INSPIRED ENGINEERING
The first Porsche 356 was a triumph of creative expediency and inspired engineering, taking basic VW Beetle elements to create a new breed of sports car. Aficionados adore the earliest cars, often affectionately dubbed "jelly moulds".

ACCESS COVER
Not a covered jacking point but an access cover to allow you to retrieve the torsion bar.

CABIN
Seats were wide and flat, and the large, almost vertical, steering wheel had a light feel. Passengers got a grab handle.

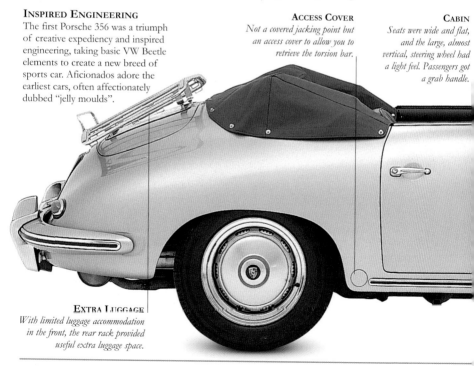

EXTRA LUGGAGE
With limited luggage accommodation in the front, the rear rack provided useful extra luggage space.

CARRERA OPTION
*The '62 356
Carrera 2 model
had a 1966cc engine.*

RACE WINNERS
The first Porsche 356s distinguished themselves
almost immediately with a 1951 Le Mans class win
and a placing of 20th overall. Since then, Porsche
have always been associated with performance,
boasting an enviable track and rally victory tally.

BRAKES
*Drum brakes gave
way to all-round
discs with the
356C in 1963.*

GEARING
*The patented Porsche baulk-
ring synchromesh gave smooth
gear changes with quick and
positive engagement.*

WHEELBASE
*The 356's
wheelbase measured
210 cm (82 in).*

'62 BLOCK
*This is the 1582cc
engine of the
1962 356B.*

SPLIT-SCREEN DECEIT
*On convertibles, the rear-
view mirror was attached to
a slim chrome bar that gave
a deceptive split-screen
appearance from the front.*

ENGINE

The rear-engined layout was determined by reliance on
VW Beetle mechanicals and running gear. The flat-four
engine, with its so-called "boxer" layout of horizontally
opposed cylinders, is not pure Beetle, but a progressive
development. Engines grew from 1086cc to 1996cc.

REDESIGN
*On the 356B, headlamps
and bumpers moved
higher up the wing.*

INTERIOR

The interior is delightfully
functional, unfussy, un-faddish,
and, because of that, enduringly
fashionable. Below the
padded dash are the
classic green-on-
black instruments.

DYU

911 PRECURSOR

The original incarnation of the 356 had lower wheels and a more bulbous shape. The featured car here is a 1962 356B Super 90, produced just two years before the birth of the 911 *(see pages 420–21)* which, although a very different beast, is still an evolution of the original shape.

REAR VIEW

On the 356B twin exhausts exit on each side through bumper over-riders. The busy air-cooled thrum is an unmistakable trademark sound that was appreciated by thousands of buyers.

SPECIFICATIONS

MODEL Porsche 356B (1959–63)

PRODUCTION 30,963

BODY STYLES Two-plus-two fixed-head coupé, cabriolet, and Speedster.

CONSTRUCTION Unitary steel body with integral pressed-steel platform chassis.

ENGINE Air-cooled, horizontally opposed flat-four 1582cc with twin carbs.

POWER OUTPUT 90 bhp at 5500 rpm (Super 90).

TRANSMISSION Four-speed manual, all synchromesh, rear-wheel drive.

SUSPENSION *Front:* independent, trailing arms with transverse torsion bars and anti-roll bar; *Rear:* independent, swing half-axles, radius arms, and transverse torsion bars. Telescopic shocks.

BRAKES Hydraulic drums all round.

MAXIMUM SPEED 77 km/h (110 mph)

0–60 MPH (0–96 KM/H) 10 sec

A.F.C. 10.6–12.5 km/l (30–35 mpg)

PORSCHE *Carrera 911 RS*

AN INSTANT LEGEND, THE CARRERA RS became the classic 911, and is hailed as one of the ultimate road cars of all time. With lighter body panels and stripped out interior trim, the RS is simply a featherweight racer. The classic, flat-six engine was bored out to 2.7 litres and boasted uprated fuel injection and forged flat-top pistons – modifications that helped to push out a sparkling 210 bhp. Porsche had no problem selling all the RSs it could make, and a total of 1,580 were built and sold in just 12 months. Standard 911s were often criticized for tail-happy handling, but the Carrera RS is a supremely balanced machine. Its race-bred responses offer the last word in sensory gratification. With one of the best engines ever made, an outstanding chassis, and 243 km/h (150 mph) top speed, the RS can rub bumpers with the world's finest. Collectors and Porsche buffs consider this the pre-eminent 911, with prices reflecting its cult-like status. The RS is the original air-cooled screamer.

LIGHTWEIGHT COUPÉ

The polyester bumpers, thin steel bodywork, and lightweight "Glaverbell" glass help the RS to weigh in at just over 900 kg (1,984 lb). Standard Porsches tip the scales at 995 kg (2,194 lb). In addition, the weight distribution and rear engine layout demand some very gentle treatment of the throttle. Handle the 911 roughly and it will understeer.

WINDSCREEN
Steeply raked screen helped the 911's wind-cheating shape.

LIGHTS
Classic slanted headlights betrayed the 911's VW Beetle origins.

REAR-ENGINED
The bored-out, air-cooled 2.7-litre
"Boxermotor" produces huge reserves
of power. Externally, it is identifiable only
by extra cylinder cooling fins.

SPECIFICATIONS

MODEL Porsche Carrera 911 RS (1972–73)

PRODUCTION 1,580

BODY STYLE Two door, two seater coupé.

CONSTRUCTION Thin-gauge steel panels.

ENGINE Flat-six, 2687cc.

POWER OUTPUT 210 bhp at 5100 rpm.

TRANSMISSION Close-ratio, five-speed manual.

SUSPENSION Front and rear torsion bar.

BRAKES Ventilated discs front and rear, with aluminium calipers.

MAXIMUM SPEED 243 km/h (150 mph)

0–60 MPH (0–96 KM/H) 5.6 sec

0–100 MPH (0–161 KM/H) 12.8 sec

A.F.C. 8.1 km/l (23 mpg)

WHEELARCHES
*Rear wheelarches were
flared to accommodate
18-cm (7-in) rims.*

REAR SPOILER
*The RS had a glass-fibre
Burzel rear spoiler, fitted to
reduce tail-end lift at speed.*

RANGE ROVER

DESCRIBING THE RANGE ROVER AS THE BEST CAR in the world is no exaggeration. The sheer breadth of the capabilities of the third-generation Rangie (as it is affectionately known) was truly awesome. Developed by BMW in the late '90s, it set new SUV standards with air suspension, voice-activated satellite navigation, the heave of a hot hatch, and the mountain-climbing tenacity of Sherpa Tenzing. The most expensive and popular Range Rover ever, the L322 was a 4x4 that felt like a Bentley and was the car that helped make Jaguar Land Rover one of the most admired and innovative car companies on the planet.

THE FRUGAL 4x4
4.4 TDV8 versions could better 12.7 km/l (30 mpg).

MAGIC CARPET
On-road ride was serenely smooth.

INTERIOR

Lush leather, cooled seats, a heated steering wheel, touch-screen TV, virtual instruments, and an eight-speed automatic gearbox all came as standard. The interior on top Autobiography models was as palatial as a Rolls Royce.

BIG SCREEN
Massive heated screen had automatic rain-sensitive wipers.

SPECIFICATIONS

MODEL Range Rover (2002–12)
PRODUCTION More than 200,000
BODY STYLE Five-door SUV.
CONSTRUCTION Monocoque.
ENGINE 3.0–5.0-litre, straight-six V8.
POWER OUTPUT 286–503 bhp.
TRANSMISSION Five- to eight-speed automatic.
SUSPENSION Independent/air.
BRAKES Four-wheel discs.
MAXIMUM SPEED 209 km/h (130 mph) (supercharged)
0–60 MPH (0–96 KM/H) 6.5 sec (supercharged)
0–100 MPH (0–161 KM/H) 14.2 sec (supercharged)
A.F.C. 8.0–12.7 km/l (19–30 mpg)

STYLING FLOURISH
Decorative side grilles broke up the huge slab-sided flanks.

A SUPERCHARGED SUV

Supercharged versions gave fierce performance and the title of "The Fastest 4x4 by Far". Jaguar-sourced alloy V8s were 4.2 litre at first, giving 395 bhp, and later enlarged to 5.0 litre, pushing out over 500 horsepower. Different grille and side vents told everyone you had a supercharger up front.

BIG BRAKES
Brakes were Brembo four-wheel discs.

RENAULT-*Alpine A110 Berlinette*

THE RENAULT-ALPINE A110 may be diminutive in its proportions but it has a massive and deserved reputation, particularly in its native France. Although wearing the Renault badge, this pocket rocket is a testimony to the single-minded dedication of one man – Jean Redélé, a passionate motor sport enthusiast and son of a Dieppe Renault agent. As he took over his father's garage he began to modify Renault products for competition, then develop his own machines based on Renault engines and mechanicals. The A110, with its glass-fibre body and backbone chassis, was the culmination of his effort, and from its launch in 1963 it went on to rack up a massive list of victories in the world's toughest rallies. On the public roads, it had all the appeal of a thinly disguised racer, as nimble as a mountain goat, with sparkling performance and just about the most fun you could have this side of a Lancia Stratos *(see pages 330–33).*

MEAN MACHINE
Squat, nimble, and slightly splay-footed on its wide tyres, the Alpine looks purposeful from any angle. Climb into that tight cockpit and you soon feel part of the car; start it up and there is a delicious barrage of noise. On the move, the sting in the Alpine's tail is exhilarating as it buzzes behind you like an angry insect.

GT4 OPTION
A short-lived 2+2 version never had the sporting attraction of the Berlinette.

COMPACT SIZE
It is a compact little package just 1.16 m (44.5 in) high, 1.5 m (60 in) wide, and 3.85 m (151.5 in) in length.

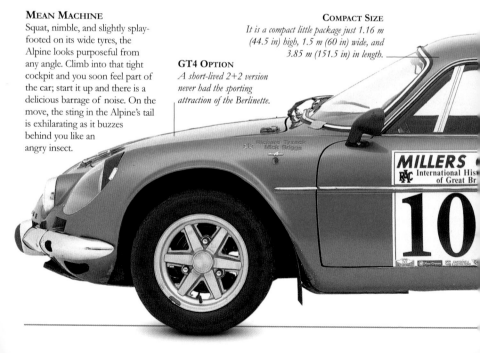

Richard Tyzack
Mick Briggs

MILLERS
International His
of Great Br

10

GO-KART HANDLING
The steering is light and the grip limpet-like, but when it does let go that tail wags the dog in a big way. Its singular appearance remained intact through its production life, with only detail changes to the trim, which these days is rare.

SPECIFICATIONS

MODEL Renault-Alpine A110 Berlinette (1963–77)

PRODUCTION 8,203

BODY STYLE Two-seater sports coupé.

CONSTRUCTION Glass-fibre body integral with tubular steel backbone chassis.

ENGINES Various four-cylinders of 956 to 1796cc.

POWER OUTPUT 51–66 bhp (956cc) to 170 bhp (1796cc)

TRANSMISSION Four- and five-speed manual, rear-wheel drive.

SUSPENSION Coil springs all round. *Front:* upper/lower control arms; *Rear:* trailing radius arms & swing-axles.

BRAKES Four-wheel discs.

MAXIMUM SPEED 212 km/h (132 mph) (1595cc)

0–60 MPH (0–96 KM/H) 8.7 sec (1255cc), 10 sec (1442cc)

A.F.C. 7.6 km/l (27 mpg) (1296cc)

BOOT AJAR
Competition versions had engine covers fixed slightly open to aid cooling.

ENGINE

Myriad engine options mirrored Renault's offerings but, in Alpine tune – by Gordini or Mignotet – it really flew. First models used Dauphine engines, progressing through R8 and R16 to R12. This 1967 car sports the 1442cc unit. Engines were slung behind the rear axle, with drive taken to the gearbox in front of the axle.

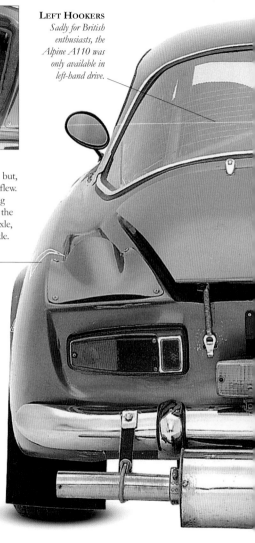

LEFT HOOKERS
Sadly for British enthusiasts, the Alpine A110 was only available in left-hand drive.

RALLY SUCCESSES
Among the myriad rally successes for Alpine were two Monte Carlo victories and the 1973 World Championship.

EXTERNAL CUT-OUT
External cut-out switches are a competition requirement, allowing outsiders to switch off the engine to prevent fire in an accident. The Alpine's are on the rear wing.

INSIDE THE CAR

Instrument layout is typical of sporting cars of the period, and the stubby gear-lever is handily placed for ease of operation. Examples built for road rather than race use lacked the racing seats but were better trimmed and were still fun cars to drive. Getting in and out was not easy though, because of the low roof line and high sills.

NAME

Cars were known at first as Alpine-Renaults, then became Renault-Alpines as Renault influence grew.

ASSEMBLY

Even though only a little over 8,000 A110s were built, they were assembled in Spain, Mexico, Brazil, and Bulgaria, as well as France.

DEALER OPTION

Alpines were sold through Renault dealers – with Renault warranty – from 1969 onwards.

ROLLS-ROYCE *Silver Cloud III*

IN 1965, £5,500 BOUGHT A seven-bedroomed house, 11 Austin Minis, or a
Rolls-Royce Silver Cloud. The Rolls that everybody remembers was the ultimate
conveyance of landed gentry and captains of industry. But, by the early Sixties,
Britain's social fabric was shifting. Princess Margaret announced she was to marry
a divorcé and aristocrats were so short of old money that they had to sell their
mansions to celebrities and entrepreneurs. Against such social revolution the
Cloud was a resplendent anachronism. Each took three months to build, weighed
two tonnes, and had 12 coats of paint. The body sat on a mighty chassis and drum
brakes were preferred because discs made a vulgar squealing noise. Beneath the
bonnet slumbered straight-six or V8 engines, whose power output was never
declared, but merely described as "sufficient". The Silver Cloud stands as a
splendid monument to an old order of breeding and privilege.

MODEL HISTORY

The Cloud I was launched in 1955 and
survived until the end of the decade, when
Rolls exchanged their six for a V8 and made
power steering standard. Cloud IIs ran until
1962, when the car enjoyed its first major
facelift – a lowered bonnet line and
the fitment of voguish
twin headlamps.

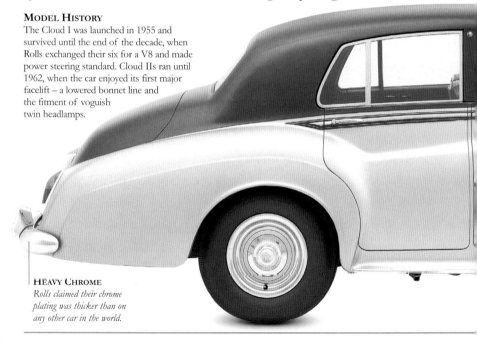

HEAVY CHROME
*Rolls claimed their chrome
plating was thicker than on
any other car in the world.*

SPECIFICATIONS

MODEL Rolls-Royce Silver Cloud III (1962–65)

PRODUCTION 2,044 Standard Steel

BODY STYLE Five-seater, four-door saloon.

CONSTRUCTION Girder chassis with pressed-steel body.

ENGINE 6230cc five-bearing V8.

POWER OUTPUT 220 bhp (estimate).

TRANSMISSION Four-speed automatic.

SUSPENSION Independent front with coils and wishbones, rear leaf springs and hydraulic dampers.

BRAKES Front and rear drums with mechanical servo.

MAXIMUM SPEED 187 km/h (116 mph)

0–60 MPH (0–96 KM/H) 10.8 sec

0–100 MPH (0–161 KM/H) 34.2 sec

A.F.C. 4.4 km/l (12.3 mpg)

INTERIOR

A haven of peace in a troubled world, the Silver Cloud's magnificent interior was a veritable throne room, with only the finest walnut, hide, and Wilton carpeting. The gear selector sat behind the steering wheel.

ENGINE

Cloud IIs and IIIs – aimed at the American market – had a 6230cc five-bearing V8 power unit, squeezed into a cramped engine bay.

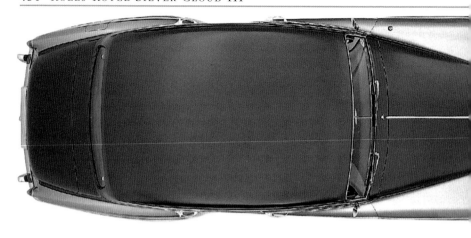

MAX HEADROOM
The roof line was high in the best limousine tradition – passengers had enough room to wear top hats. The wide rear three-quarter panel was designed so rear occupants could be obscured from prying eyes.

TOP SECURITY
Doors were secured by the highest quality Yale locks.

TOOLKIT
Every Cloud had a complete toolkit in the boot.

SCRIPT
Roman numerals were chosen for the Cloud III script to lend an air of dignity.

ANTIQUE STYLING

Everything about the Cloud's styling was antique, looking more like a piece of architecture than a motor car. Standard steel bodies were made by the Pressed Steel Co. of Oxford, England, with the doors, bonnet, and boot lid hand-finished in aluminium to save weight.

LEATHER COMFORT

The rear compartment might have looked accommodating, but Austin's little 1100 actually had more legroom. Standard walnut picnic tables were ideal for Champagne and caviar picnics. Rear leaf springs and hydraulic dampers kept the ride smooth.

FRONT ASPECT

The 150-watt 14-cm (5¼-in) Lucas double headlamps were necessitated by onerous North American safety requirements. Turn indicators were moved from the fog light to the front wing on the Cloud III.

MASCOT

The Spirit of Ecstasy graced a silver radiator shell that took several men five hours to polish.

ROLLS ROYCE *Phantom Drophead*

ROLLS ROYCE CONVERTIBLES have always been the choice of high rollers, but when the first Phantom drop-top was auctioned, it sold for four times its list price. Like the Corniche convertible before it, this is one of the world's most expensive and desirable rag-tops. Despite weighing nearly three tonnes, it can whisper to 96 km/h (60 mph) in 5.7 seconds thanks to an all-alloy construction and a 435 bhp V12 engine. In fact the sheer speed and fingertip agility of the Phantom are what make it totally unique. Hand-made in the RR factory in Goodwood, and available in 44,000 different colours, it's a gorgeous mix of art deco and techno modern. Several examples were used in the 2012 London Olympics closing ceremony. No car is more British than the Phantom and no convertible is more rock'n'roll.

REAR DECK
The rear tonneau cover is made from wood and echoes the nautical wood colours used in Italian Riva motor launches of the '50s and '60s. The interior also has acres of timber, available in hundreds of different colours and finishes.

DETAIL PERFECTION
Cabin is full of leather, chrome, alloy, wood, and crystal.

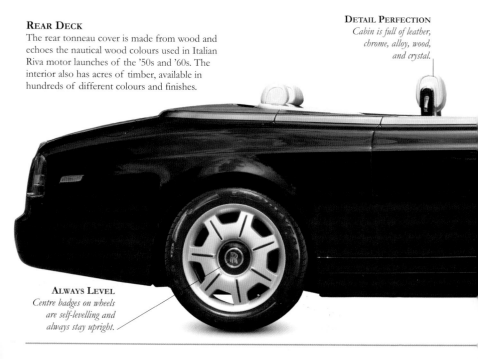

ALWAYS LEVEL
Centre badges on wheels are self-levelling and always stay upright.

MOVING MASCOT
*Spirit of Ecstasy falls
and rises automatically
when the car is locked
or unlocked.*

SPECIFICATIONS

MODEL Rolls Royce Phantom
Drophead (2007)
PRODUCTION N/A
BODY STYLE Two-door, four-seater
convertible.
CONSTRUCTION All-alloy.
ENGINE 6,749cc V12.
POWER OUTPUT 435 bhp.
TRANSMISSION Six-speed automatic.
SUSPENSION Self-levelling air suspension.
BRAKES Four-wheel discs.
MAXIMUM SPEED 233 km/h (145 mph)
(limited)
0–60 MPH (0–96 KM/H) 5.7 sec
0–100 MPH (0–161 KM/H) 14 sec
A.F.C. 6.4 km/l (15 mpg)

ROLLS REINVENTED

When BMW bought Rolls Royce, many
experts believed the brand was beyond
saving. The German firm's reinvention of
RR has made it the most powerful and
desirable it has ever been, attracting a much
more fashionable and younger customer.

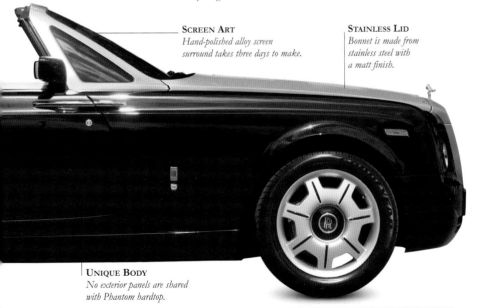

SCREEN ART
*Hand-polished alloy screen
surround takes three days to make.*

STAINLESS LID
*Bonnet is made from
stainless steel with
a matt finish.*

UNIQUE BODY
*No exterior panels are shared
with Phantom hardtop.*

Saab *99 Turbo*

Every decade or so, one car comes along that overhauls accepted wisdom. In 1978, the British motoring magazine *Autocar* wrote, "this car was so unpredictably thrilling that the adrenalin started to course again, even in our hardened arteries". They had just road-tested a Saab 99 Turbo. Saab took all other car manufacturers by surprise when they announced the world's first turbocharged family car, which promptly went on to be the first "blown" car to win a World Championship rally. Developed from the fuel-injected EMS model, the Turbo had Bosch K-Jetronic fuel injection, a strengthened gearbox, and a Garrett turbocharger. A hundred prototypes were built, and between them they covered 4.8 million kilometres (2.9 million miles) before Saab were happy with their prodigy. Although it was expensive, there was nothing to equal its urge. Rare, esoteric, and historically significant, the mould-breaking 99 Turbo is an undisputed card-carrying classic.

OFFICIAL PRESENCE
The body has a certain business-like presence, helped by specially made Inca alloys designed to mimic the shape of turbocharger blades, front and rear spoilers, and a sliding steel sunroof.

INTERIOR
Seventies interior looks a mite tacky now, with red velour seats and imitation wood.

SUSPENSION
Dead beam-axle at the rear and the usual wishbone and coil spring set-up at the front.

ENGINE
The five-bearing, chain-driven single overhead cam
engine was an 1985cc eight-valve, water-cooled,
four cylinder unit, with low-compression pistons.

SPECIFICATIONS

MODEL Saab 99 Turbo (1978–80)

PRODUCTION 10,607

BODY STYLES Two/three/five-door,
four-seater sports saloon.

CONSTRUCTION Monocoque steel
bodyshell.

ENGINE 1985cc four-cylinder turbo.

POWER OUTPUT 145 bhp at 5000 rpm.

TRANSMISSION Front-wheel drive
four/five-speed manual with auto option.

SUSPENSION Independent front double
wishbone and coil springs, rear beam axle,
coil springs, and Bilstein shock absorbers.

BRAKES Four-wheel servo discs.

MAXIMUM SPEED 196 km/h (122 mph)

0–60 MPH (0–96 KM/H) 8.2 sec

0–100 MPH (0–161 KM/H) 19.8 sec

A.F.C. 9.3 km/l (26 mpg)

TURBOCHARGER
*The turbo was reliable, but its
Achilles' heel was a couple of
seconds' lag on hard acceleration.*

HANDLING
*99 Turbos were poised.
Crisp turn-in came from
front-wheel drive, with
prodigious adhesion courtesy
of 195/60 Pirelli P6s.*

STUDEBAKER *Avanti*

THE AVANTI WAS A BIG DEAL for Studebaker and the first all-new body style since 1953. The last car design of the legendary Raymond Loewy, it rode on a shortened Lark chassis with a stock Studey 289cid V8. The Avanti's striking simplicity of shape was just one of Loewy's celebrated confections. From his voguish Coca-Cola dispenser to the chaste Lucky Strike cigarette packet, Loewy's creations were instant classics, and the brilliant Avanti was a humdinger. Studebaker's prodigy was fairly audacious too, with a glass-fibre body, anti-sway bars, and wind-cheating aerodynamics. Dealers, however, could not meet the huge wave of orders and this, combined with other niggles like flexing of the glass-fibre shell, resulted in impatient buyers defecting to the Corvette camp instead. Fewer than 4,650 Avantis were made, and production ceased in December 1963, the Avanti concept being sold to a couple of Studebaker dealers. They went on to form the Avanti Motor Corporation, which successfully churned out Avantis well into the Eighties.

EUROPEAN LINES
More European than American, the Avanti had a long neck, razor-edged front wings, and no grille. Early sketches show Loewy's inspiration, with tell-tale annotations scribbled on the paper that read "like Jaguar, Ferrari, Aston Martin, Mercedes". Lead time for the show Avanti was a hair-raising 13 months, with a full-scale clay model fashioned in only 40 days.

ENGINE
The 289cid was the best Studebaker V8 ever made, developing 240 bhp in standard R1 tune. Supercharged R2 and R3 boasted 290 and 335 bhp respectively.

BODY STYLING
The slippery shape was not wind-tunnel tested, but a piece of guesswork by Loewy.

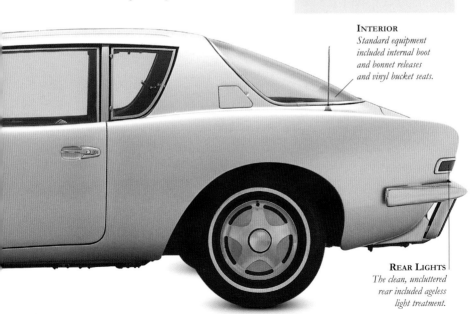

FRONT VIEW
Unmistakable from any angle, early '63
Avantis had round headlights, but most later
'64 models sported square ones.

SPECIFICATIONS

MODEL Studebaker Avanti (1963)

PRODUCTION 3,834 (1963)

BODY STYLE Two-door, four-seater coupé.

CONSTRUCTION Glass-fibre body,
steel chassis.

ENGINES 289cid, 304cid V8s.

POWER OUTPUT 240–575 bhp
(304cid R5 V8 fuel-injected).

TRANSMISSION Three-speed manual,
optional Power-Shift automatic.

SUSPENSION *Front:* upper and lower
A-arms, coil springs; *Rear:* leaf springs.

BRAKES Front discs, rear drums.

MAXIMUM SPEED 193 km/h
(120 mph)

0–60 MPH (0–96 KM/H) 7.5 sec

A.F.C. 6 km/l (17 mpg)

INTERIOR
*Standard equipment
included internal boot
and bonnet releases
and vinyl bucket seats.*

REAR LIGHTS
*The clean, uncluttered
rear included ageless
light treatment.*

SUNBEAM *Tiger*

THERE WAS NOTHING NEW ABOUT popping an American V8 into a pert English chassis. After all, that is exactly what Carroll Shelby did with the AC Ace to create the awesome Cobra *(see pages 16–19)*. When Rootes in Britain decided to do the same with their Sunbeam Alpine, they also commissioned Shelby to produce a prototype, and although Rootes already had close links with Chrysler, the American once again opted for a Ford V8. To cope with the 4.2-litre V8, the Alpine's chassis and suspension were beefed up to create the fearsome Tiger late in 1964. In 1967, the Tiger II arrived with an even bigger 4.7-litre Ford V8, but this was a brief swansong as Chrysler took control of Rootes and were not going to sanction a car powered by rivals Ford. Once dubbed "the poor man's Cobra", these days Tiger prices are only for the rich.

ENGINE
The first Tigers used 4.2-litre Ford V8 engines, replaced later – as shown here – by a 4727cc version, the famous 289, but not in the same state of tune as those fitted to the Shelby Cobras.

DISTINGUISHING FEATURES
The MkII Tiger had an egg-crate grille to distinguish it from the Alpine. Earlier cars were less easy to tell apart: a chrome strip along the side of the Tiger was the giveaway, together with discreet badging on the body.

ADAPTING THE ALPINE
The Alpine's chassis and suspension had to be beefed up to cope with the weight and power of the V8. Resulting modifications included a heavy-duty back axle, sturdier suspension, and chassis stiffening.

RACE BONNET
Race and rally Tigers had improved air-flow with a slightly raised bonnet.

HOT HOUSE
Tigers often suffered from overheating.

SPECIFICATIONS

MODEL Sunbeam Tiger (1964–67)

PRODUCTION 6,496 (Mk1, 1964–67); 571 (MkII).

BODY STYLE Two-plus-two roadster.

CONSTRUCTION Steel monocoque.

ENGINES Ford V8 4261cc or 4727cc (260 or 289cid).

POWER OUTPUT 164 bhp at 4400 rpm (4261cc), 200 bhp at 4400 rpm (4727cc).

TRANSMISSION Four-speed manual.

SUSPENSION Coil springs and wishbones at front, rigid axle on semi-elliptic leaf springs at rear.

BRAKES Servo-assisted front discs, rear drums.

MAXIMUM SPEED 188 km/h (117 mph) (4261cc), 201 km/h (125 mph) (4727cc)

0–60 MPH (0–96 KM/H) 9 sec (4261cc), 7.5 sec (4727cc).

A.F.C. 7 km/l (20 mpg)

Tesla *Roadster*

THE TESLA WAS THE WORLD'S FIRST sexy electric car. Fast enough to worry a Porsche 911 Turbo or Ferrari 599, the neck-jerking torque and devastating, silent acceleration felt uncanny. Brainchild of PayPal founder Elon Musk, it used a Lotus Elise chassis, stored power in 6,800 laptop batteries, and was the first electric vehicle (EV) to have a range of over 322 km (200 miles) on a three-and-a-half-hour charge. With zero tailpipe emissions and a theoretical fuel consumption of 51 km/l (120 mpg), its green credentials were unimpeachable, but it also gave the struggling EV market glamour and desirability. Without the ferociously fast Tesla, electric cars wouldn't have come as far as they have. This sparky little roadster remains one of the great technological landmark cars of the 21st century.

SMOOTH FEATHERWEIGHT
Slippery aerodynamics and lightweight construction make it as fast as a Lamborghini Gallardo.

ENVIABLE RANGE
The biggest barrier to volume electric car sales has always been "range anxiety", or the fear of running out of battery power. But in 2010 Tesla cleverly and successfully drove a roadster round the world, and owners often achieved over 482 km (300 miles) to one charge.

LOTUS BACKBONE
The Lotus Elise chassis and suspension gives scalpel-sharp handling.

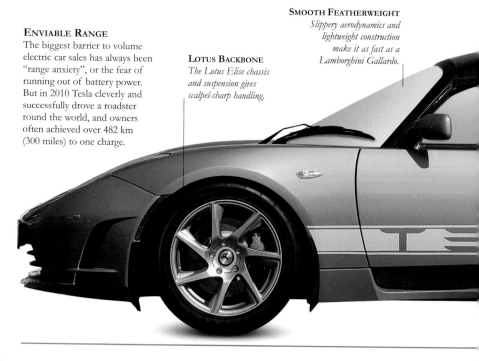

QUICK POWER
Fast charging system is twice as quick as most other EVs.

COOL RIDE
Wing ducts help cool batteries and brakes.

EASY ELECTRICITY
Owners have a home charging unit that simply plugged in, and can also top up at the office or at a street charger, if they can find one.

ROOFLESS
Removable roof panel makes a quick convertible.

STOPPING POWER
Drilled, ventilated four-wheel discs are extremely powerful.

SPECIFICATIONS

MODEL Tesla Roadster (2007)
PRODUCTION 2,450
BODY STYLE Two-door, two-seater.
CONSTRUCTION Steel chassis, carbon-fibre panels.
ENGINE 185-kw electric motor.
POWER OUTPUT 248 hp.
TRANSMISSION Single-speed Borg Warner.
SUSPENSION Independent.
BRAKES Four-wheel discs.
MAXIMUM SPEED 53 km/h (125 mph) (limited)
0–60 MPH (0–96 KM/H) 3.9 sec
0–100 MPH (0–161 KM/H) 12 sec
A.F.C. Theoretical 51 km/l (120 mpg)

TOYOTA *2000GT*

TOYOTA'S 2000GT IS MORE than a "might have been" – it's a "should have been". A pretty coupé with performance and equipment to match its good looks, it pre-dated the rival Datsun 240Z *(see pages 196–99)*, which was a worldwide sales success. The Toyota failed to reach much more than 300 sales partly because of low capacity, but even more because the car was launched before Japan was geared to export. That left only a domestic market, largely uneducated in the finer qualities of sporting cars, to make what they could of the offering. As a design exercise, the 2000GT proved that the Japanese motor industry had reached the stage where its products rivalled the best in the world. It is just a pity not more people were able to appreciate this fine car at first hand.

BEEMER LINKS

The design of the Toyota 2000GT is based on an earlier prototype penned by Albrecht Goertz, creator of the BMW 507 *(see pages 64–67)* and Datsun 240Z. When Nissan rejected the design, it was offered to Toyota and evolved into the 2000GT.

GEAR LEVER
Short-throw wooden-top gear lever.

INTERIOR

The 2000GT's snug cockpit featured a walnut-veneer instrument panel, sporty wheel, stubby gear-lever, form-fitting seats, and deep footwells. The eight-track stereo is a nice period touch.

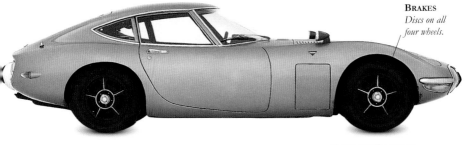

BRAKES
*Discs on all
four wheels.*

LIGHTING
*Unusual combination
of high-tech pop-up and
fixed headlights gave
the front a fussy look.*

BONNET PROFILE

The panel on the right concealed the
GT's battery; the one on the left-hand
side of the body was the air cleaner.
This arrangement enabled the bonnet
to be kept low. The engine was a triple-
carb six-cylinder Yamaha, which
provided 150 bhp. A competition
version boosted output to 200 bhp.

SPECIFICATIONS

MODEL Toyota 2000GT (1966–70)

PRODUCTION 337

BODY STYLE Two-door sports coupé.

CONSTRUCTION Steel body on
backbone frame.

ENGINE Yamaha inline DOHC six, 1988cc.

POWER OUTPUT 150 bhp at 6600 rpm.

TRANSMISSION Five-speed manual.

SUSPENSION Fully independent by coil
springs and wishbones all round.

BRAKES Hydraulically operated discs
all round.

MAXIMUM SPEED 206 km/h (128 mph)

0–60 MPH (0–96 KM/H) 10.5 sec

0–100 MPH (0–161 KM/H) 24 sec

A.F.C. 11 km/l (31 mpg)

TRIUMPH *TR2*

IF EVER THERE WAS A SPORTS CAR that epitomized the British bulldog spirit it must be the Triumph TR2. It is as true Brit as a car can be, born in the golden age of British sports cars, but aimed at the lucrative American market. At the 1952 Earl's Court Motor Show in London, the new Austin-Healey stole the show, but the "Triumph Sports" prototype's debut at the same show was less auspicious. It was a brave attempt to create an inexpensive sports car from a company with no recent track record in this market segment. With its dumpy *derriere*, the prototype was no oil painting; as for handling, chief tester Ken Richardson described it as a "bloody deathtrap". No conventional beauty certainly, but a bluff-fronted car that was a worthy best-of-breed contender in the budget sports car arena, and the cornerstone of a stout sporting tradition.

UNCONVENTIONAL STYLING
The design, by Walter Belgrove, was a far cry from the razor-edged Triumph Renown and Mayflower saloons that he had previously styled. If not beautiful, the TR2 has chunky good looks with a bluff, honest demeanour.

HOOD
The TR2 had a foldaway hood; the later TR3 had the option of a lift-off hardtop.

RACING HOLES
The TR2 came with small holes drilled in the scuttle to fit aero-screens for racing.

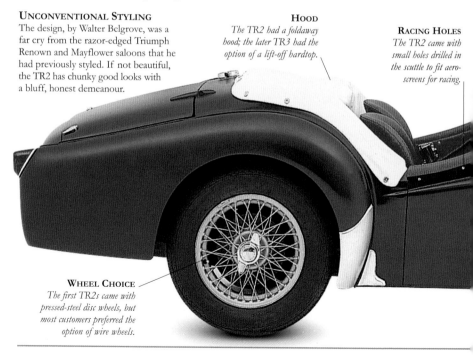

WHEEL CHOICE
The first TR2s came with pressed-steel disc wheels, but most customers preferred the option of wire wheels.

OVERHEAD VIEW
The low-cut doors meant that you could
reach out over them and touch the road.
External door handles only arrived
with the TR3A of 1957.

CHASSIS
*The TR2 chassis was
praised for its tautness
and fine road manners.*

FUEL FILLER
*At over 10.6 km/l
(30 mpg), the TR2's fuel
figures were impressive.*

WINDSCREEN
The windscreen had a slight curve to prevent it from bowing at speed, which is what the prototype's flat screen did.

SPORTING SUCCESS
TR2s came first and second in the 1954 RAC Rally.

NEW REAR
A revised rear, all-new chassis, and other modifications saw Standard-Triumph's new TR2 emerge into a winner at the Geneva Motor Show in March 1953. While the prototype had a stubby tail, the production model had a real opening boot.

SPECIFICATIONS

MODEL Triumph TR2 (1953–55)

PRODUCTION 8,628

BODY STYLE Two-door, two-seater sports car.

CONSTRUCTION Pressed-steel chassis with separate steel body.

ENGINE Four-cylinder, overhead valve, 1991cc, twin SU carburettors.

POWER OUTPUT 90 bhp at 4800 rpm.

TRANSMISSION Four-speed manual with Laycock overdrive option, initially on top gear only, then on top three (1955).

SUSPENSION Coil-spring and wishbone at front, live rear axle with semi-elliptic leaf springs.

BRAKES Lockheed hydraulic drums.

MAXIMUM SPEED 169 km/h (105 mph)

0–60 MPH (0–96 KM/H) 12 sec

A.F.C. 10.6+ km/l (30+ mpg)

FRONT VIEW

The unusual recessed grille perhaps presents a slightly grumpy disposition, but the low front helped the car to a top speed of 169 km/h (105 mph). Fittings on the TR2 were spartan – you did not even get external door-handles.

AXLE

Like the Austin-Healey's, the TR chassis ran under the axle at the rear.

STOCK DESIGN

There is nothing revolutionary in the design of the pressed-steel chassis; a simple ladder with X-shaped bracing. It was a transformation, though, from the prototype's original chassis.

INTERIOR

Stubby gear lever and full instrumentation gave TR a true sports car feel; the steering wheel was large, but the low door accommodated "elbows out" driving style.

Triumph *TR6*

To most TR traditionalists this is where the TR tale ended, the final flourishing of the theme before the TR7 betrayed an outstanding tradition. In the mid-Sixties, the TR range was on a roll and the TR6 continued the upward momentum, outselling all earlier offerings. It was a natural progression from the original TR2; the body evolved from the TR4/5, the power unit from the TR5. Crisply styled, with chisel-chin good looks and carrying over the 2.5-litre six-cylinder engine of the TR5, the TR6 in early fuel-injected form heaved you along with 152 galloping horses. This was as hairy chested as the TR got, and a handful too, with some critics carping that, like the big Healeys, its power outstripped its poise. But that just made it more fun to drive.

Karmann Styling

There is an obvious difference between the TR4/5 and the later TR6, restyled by Karmann; sharper, cleaner lines not only looked more modern, but also gave more luggage space. The chopped off tail was an aerodynamic aid.

Top Option

One-piece hardtop was available as an option, and more practical than the two-piece job seen on earlier models.

Stateside Sales

Some 78,000 TR6s went to the US even though emission regulations emasculated it.

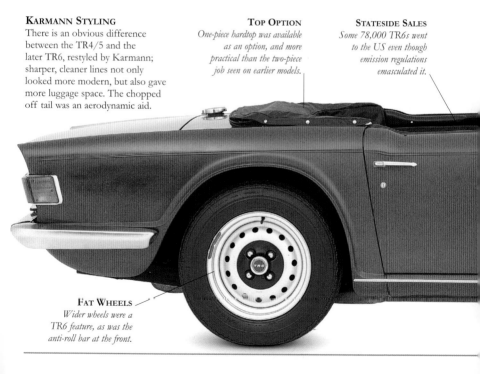

Fat Wheels
Wider wheels were a TR6 feature, as was the anti-roll bar at the front.

SMOOTH TR6
Virtually all bulges, like the TR5's bonnet "power bulge" and cowled headlights, have been ironed out.

ROOMY COCKPIT
The cockpit was more spacious than earlier TRs, providing excellent driving position from comfortable seats. Big, wide-opening doors gave easy access to the TR6, a long cry from the tiny doors of the TR2 and 3.

POWER DROP
Revised injection metering and reprofiled camshaft reduced power from 1973; US carburettor versions were more sluggish and thirstier.

BEST SELLER

The TR6's good looks, and a long production run, made this model the biggest selling of all TR models. British sales stopped in February 1975, but continued in the US until July 1976. The US model may have been slower than the UK model by 19 km/h (12 mph), but 10 times as many TR6s were exported as remained in Britain.

ENGINE

The first engines, as on this 1972 car, produced 152 bhp, but public pressure for something more well-mannered resulted in a 125 bhp version in 1973. Americans had to make do with just over 100 bhp and no fuel injection.

STEERING WHEEL

Steering-wheel size was reduced at the time of other mid-model changes in 1973.

INTERIOR

The interior is still traditional but more refined than earlier TRs. Yet with its big dials, wooden fascia, and short-throw gear knob, its character is still truly sporting.

MERGER

The TR6 was launched just after the 1968 merger of Leyland and BMC, which produced Triumph motors. Hence the badge on the side of the TR6's bodywork.

ENGINE NOISE

Deep-throated burble is still a TR6 come-on.

LONG-TAILED

The TR6's squared-off tail was longer than earlier TRs. Even so, there was only space in the boot for a set of golf clubs and an overnight bag.

SPECIFICATIONS

MODEL Triumph TR6 (1969–76)

PRODUCTION 94,619

BODY STYLE Two-seat convertible.

CONSTRUCTION Ladder-type chassis with integral steel body.

ENGINE Inline six-cylinder, 2498cc, fuel-injection (carburettors in US).

POWER OUTPUT 152 bhp at 5500 rpm (1969–1973), 125 bhp at 5250 rpm (1973–1975), 104 bhp at 4500 rpm (US).

TRANSMISSION Manual four-speed with optional overdrive on third and top.

SUSPENSION Independent by coil springs all round; wishbones at front, swing-axles & semi-trailing arms at rear.

BRAKES *Front:* discs; *Rear:* drums.

MAXIMUM SPEED 191 km/h (119 mph, 150 bhp), 172 km/h (107 mph, US)

0–60 MPH (0–96 KM/H) 8.2 sec (150 bhp); 9.0 sec (125 bhp); 10.6 sec (104 bhp)

0–100 MPH (0–161 KM/H) 29 sec

A.F.C. 8.8 km/l (25 mpg)

TUCKER *Torpedo*

THERE'S NO OTHER POST-WAR CAR that's as dramatic or advanced as Preston Tucker's futuristic '48 Torpedo. With four-wheel independent suspension, rear-mounted Bell helicopter engine, pop-out safety windscreen, and uncrushable passenger compartment, it was 20 years ahead of its time. "You'll step into a new automotive age when you drive your Tucker '48", bragged the ads. It was a promise that convinced an astonishing 300,000 people to place orders, but their dreams were never to be realized. Problems with the engine and Tuckermatic transmission, plus a serious cash-flow crisis, meant that only 51 Torpedos left the Chicago plant. Worse still, Tucker and five of his associates were indicted for fraud by the Securities Exchange Commission. Their acquittal came too late to save America's most eccentric car from an undignified end.

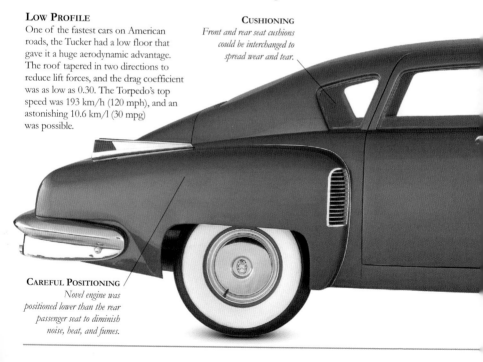

LOW PROFILE
One of the fastest cars on American roads, the Tucker had a low floor that gave it a huge aerodynamic advantage. The roof tapered in two directions to reduce lift forces, and the drag coefficient was as low as 0.30. The Torpedo's top speed was 193 km/h (120 mph), and an astonishing 10.6 km/l (30 mpg) was possible.

CUSHIONING
Front and rear seat cushions could be interchanged to spread wear and tear.

CAREFUL POSITIONING
Novel engine was positioned lower than the rear passenger seat to diminish noise, heat, and fumes.

ENGINE

The first of the Tucker engines was a monster 589cid aluminium flat-six that proved difficult to start and ran too hot. It was replaced by a 6ALV 335cid flat-six block, developed by Air-Cooled Motors of Syracuse. Perversely, Tucker later converted this unit to a water-cooled system.

SPECIFICATIONS

MODEL Tucker Torpedo (1948)

PRODUCTION 51 (total)

BODY STYLE Four-door sedan.

CONSTRUCTION Steel body and chassis.

ENGINE 335cid flat-six.

POWER OUTPUT 166 bhp.

TRANSMISSION Three-speed Tuckermatic automatic, four-speed manual.

SUSPENSION Four-wheel independent.

BRAKES Front and rear drums.

MAXIMUM SPEED 193 km/h (120 mph)

0–60 MPH (0–96 KM/H) 10.1 sec

A.F.C. 10.6 km/l (30 mpg)

INTERIOR DESIGN

Interior was designed by Audrey Moore, who had worked with Raymond Loewy on Studebakers.

LUGGAGE SPACE

With no engine upfront, luggage space was commodious.

NOSE DESIGN

Slippery front was designed to cleave the air.

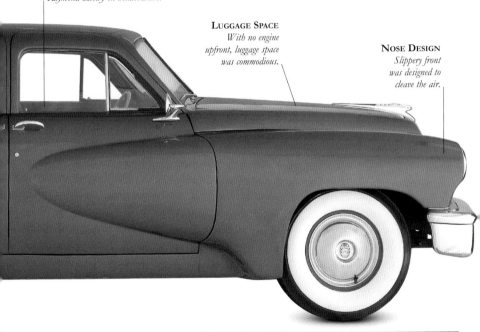

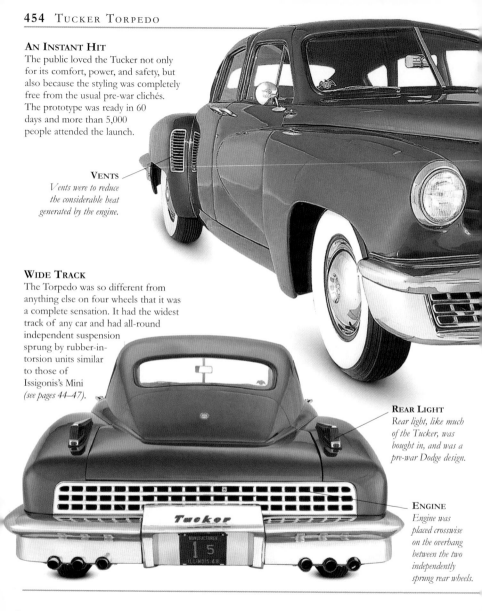

An Instant Hit
The public loved the Tucker not only for its comfort, power, and safety, but also because the styling was completely free from the usual pre-war clichés. The prototype was ready in 60 days and more than 5,000 people attended the launch.

Vents
Vents were to reduce the considerable heat generated by the engine.

Wide Track
The Torpedo was so different from anything else on four wheels that it was a complete sensation. It had the widest track of any car and had all-round independent suspension sprung by rubber-in-torsion units similar to those of Issigonis's Mini *(see pages 44–47).*

Rear Light
Rear light, like much of the Tucker, was bought in, and was a pre-war Dodge design.

Engine
Engine was placed crosswise on the overhang between the two independently sprung rear wheels.

CYCLOPS LIGHT
Daring cyclops headlight swivelled with the front wheels.

TUCKER BADGE
The horn on the steering wheel lay flush for safety and was adorned with the Tucker family crest.

INTERIOR
Some say that Detroit conspired to destroy Tucker, but steering wheels on Torpedos were from the Lincoln Zephyr, given freely by Ford as a gesture of assistance. Although the interior was groaning with safety features, the Tucker sales team reckoned it was too austere.

BUMPER
Steerhorn bumper gave the car a dramatic frontal aspect.

UNIQUE AND EXCITING
The front was like no other American car, with a fixed circular headlight lens that pivoted with the steering and a front panel that blended artfully into the bumper and grille. Designed by former Auburn-Cord-Duesenberg stylist Alex Tremulis, the Tucker was so low that it only came up to a man's shoulder.

VOLKSWAGEN *Beetle Karmann*

BEETLE PURISTS MAY WAX lyrical about the first-of-breed purity of the original split-rear-screen Bugs and the oval window versions of 1953 to 1957, but there is one Beetle that everybody wants – the Karmann-built Cabriolet. Its development followed that of the saloons through a bewildering series of modifications, but it always stood apart. With its hood retracted into a bulging bustle, this Beetle was not only cheerful, but chic too, a classless cruiser at home equally on Beverly Hills boulevards, Cannes, and the Kings Road. The final incarnation of the Karmann convertible represents the ultimate development of the Beetle theme, with the peppiest engine and improved handling. It's strange to think that the disarming, unburstable Bug was once branded with the slogan of the Hitler Youth, "Strength through Joy". Today, modern retro Beetles have become suburban middle-class trinkets.

ROADSTER PLANS

Before Karmann chopped the lid off the Bug, there had been plans for a Beetle-based roadster. The prototypes inspired coachbuilders Joseph Hebmüller & Sons to build a short-lived roadster, but just 696 were built before a factory fire scuppered the project.

SURFMOBILE
Cabriolets like this Californian-registered car are a mainstay of surfing culture.

BRAKES
Front discs were introduced in 1966.

INTERIOR

The Beetle is still bare, its dash dominated by the one minimal instrument; on this model the speedo incorporates a fuel gauge. It also has a padded dash, replacing the original metal fascia.

SPECIFICATIONS

MODEL VW Beetle Karmann Cabriolet (1972–1980)

PRODUCTION 331,847 (Karmann Cabriolets from 1949 to 1980).

BODY STYLE Four-seater cabriolet.

CONSTRUCTION Steel-bodied, separate chassis/body.

ENGINE Rear-mounted, air-cooled flat-four, 1584cc.

POWER OUTPUT 50 bhp at 4000 rpm.

TRANSMISSION Four-speed manual.

SUSPENSION *Front:* independent MacPherson strut; *Rear:* independent trailing arm and twin torsion bars.

BRAKES Front discs, rear drums.

MAXIMUM SPEED 133 km/h (82.4 mph)

0–60 MPH (0–96 KM/H) 18 sec

A.F.C. 8.5–10.6 km/l (24–30 mpg)

UNIT GROWTH
The Beetle's capacity grew from 1131cc to 1584cc; the engines have a deserved reputation as robust, rev-happy units.

Karmann Coachbuilder
As well as the Beetle convertible, Karmann also built the Type 1 VW Karmann-Ghia, a two-seater based on Beetle running gear.

Rear Lights
Many later design changes like these "elephant footprint" rear light clusters were driven by US regulations.

Fresh Air
With the hood raised, the Karmann cabriolet is a bit claustrophobic, but it comes into its own as a timeless top-down cruiser that is still a full four-seater. Rear vision with the top up is not much better than on early split-screen and oval-windowed models.

One-model Policy
The one-model policy that VW adopted in its early years was successful while Beetle sales soared, but by 1967 Fiat had overtaken VW as Europe's biggest car manufacturer. It was not until 1974 that the Golf and Polo revived the company's fortunes.

Engine
You can always tell that a Beetle is on its way before it comes into sight thanks to the distinctive buzzing of the air-cooled, horizontally opposed four-cylinder engine.

WINDSCREEN
*Curved "panoramic"
windscreen replaced the
flat screen in 1972.*

INDICATORS
*First cars had
semaphores; then
indicators were
wing-mounted.*

Volkswagen *Golf GTi*

Every decade or so a really great car comes along. In the Seventies it was the Golf. Like the Beetle before it, the Golf was designed to make inroads into world markets, yet while the Beetle evolved into the perfect consumer product, the Golf was planned that way. The idea of a "hot" Golf was not part of the grand plan. It was the brainchild of a group of enthusiastic Volkswagen engineers who worked evenings and weekends, impressing VW's board so much that the GTi became an official project in May 1975. Despite its youth, the GTi is as much of a classic as any Ferrari. Its claim to fame is that it spawned a traffic jam of imitators and brought an affordable cocktail of performance, handling, and reliability to the mass-market buyer. Few other cars have penetrated the suburban psyche as deeply as the original Golf GTi, and fewer still have had greatness thrust upon them at such an early age.

GTi Enhancements
GTi suspension was lower and firmer than the standard Golf, with wider tyres and wheels. Front disc brakes were ventilated, but keeping standard drums at the rear was a mistake – early Golfs were very disinclined to stop.

Hatchback
The Mk I Golf was the first of the Seventies' hatchbacks.

Alloys
Much admired cross-spoke BBS alloy wheels were both a factory-fitted and after-market option.

SIMPLE FRONT
Factory spec Golfs were understated, with just a GTi badge and a thin red stripe around the grille.

SPECIFICATIONS

MODEL Volkswagen Golf GTi Mk 1 (1976–83)

PRODUCTION 400,000

BODY STYLE Three-door five-seater hatchback.

CONSTRUCTION All steel/monocoque body.

ENGINES Four-cylinder 1588cc/1781cc.

POWER OUTPUT 110–112 bhp at 6100 rpm.

TRANSMISSION Four- or five-speed manual.

SUSPENSION *Front:* independent; *Rear:* semi-independent trailing arm.

BRAKES Front discs, rear drums.

MAXIMUM SPEED 179 km/h (111 mph)

0–60 MPH (0–96 KM/H) 8.7 sec

0–100 MPH (0–161 KM/H) 18.2 sec

A.F.C. 10.3 km/l (29 mpg)

ENGINE
Capable of 240,000 km (150,000 miles) in its stride, the 1588cc four-cylinder power unit breathed through Bosch K-Jetronic fuel injection.

VOLVO *P1800*

THERE HAS NEVER BEEN A VOLVO like the P1800, for this was a one-time flight of fancy by the sober Swedes, who already had a reputation for building sensible saloons. As a sports car the P1800 certainly looked stunning, every sensuous curve and lean line suggesting athletic prowess. But under that sharp exterior were most of the mechanicals of the Volvo Amazon, a worthy workhorse saloon. Consequently, the P1800 was no road-burner; it just about had the edge on the MGB *(see pages 372–73)*, but only in a straight line. Another competitor, the E-Type Jag *(see pages 306–09)*, was launched in 1961, the same year as the P1800 and at almost the same price, but there the comparison ends. The P1800 did have style, though, and its other virtues were pure Volvo – strength, durability, and reliability. These combined to create something quite singular in the sporting idiom – a practical sports car.

DESIGN CREDITS
Official Volvo history credits the award-winning design of the P1800 to Frua of Italy, but it was actually penned by young Swede Pelle Petterson, then a trainee at Ghia. The Italian influences are obvious in the final form.

JENSEN'S SIGNATURE
Early cars like this were built in Britain by Jensen.

LUGGAGE SPACE
As you would expect, the sensible sports car had a decent-sized boot.

SPECIFICATIONS

MODEL Volvo P1800 (1961–73)

PRODUCTION 47,707 (all models)

BODY STYLES Two-plus-two fixed-head coupé; sports estate (P1800ES).

CONSTRUCTION Unitary steel body/chassis.

ENGINES 1778cc straight-four, overhead valves; 1985cc from 1968–73.

POWER OUTPUT 100 bhp at 5500 rpm (P1800); 124 bhp at 6000 rpm (P1800E, P1800 ES).

TRANSMISSION Four-speed manual with overdrive/optional automatic.

SUSPENSION *Front:* independent coil-sprung with wishbones; *Rear:* rigid axle, coil-sprung, Panhard rod.

BRAKES Front discs, rear drums.

MAXIMUM SPEED 185 km/h (115 mph) (P1800 E/ES)

0–60 MPH (0–96 KM/H) 9.7–13.2 sec

0–100 MPH (0–161 KM/H) 31.4–53 sec

A.F.C. 7–10 km/l (20–25 mpg)

ENGINE
Early cars had 1778cc four-cylinder units with twin SU carbs; the 1985cc unit came later, followed by electronic fuel injection. All versions are reliable, willing revvers.

SAFETY MEASURES
The P1800 had a padded dash and seatbelts of Volvo's own design.

GEARING
Super-tough gearbox had excellent synchromesh.

WHEELS
Stylized fake spokes identify this as an early P1800.

WILLYS *Jeep MB*

AS ONE WAR CORRESPONDENT SAID, "It's as faithful as a dog, as strong as a mule, and as agile as a mountain goat". The flat-winged Willys Jeep is one of the most instantly recognizable vehicles ever made. Any American TV or movie action hero who wasn't on a horse was in a Jeep. Even General Eisenhower was impressed, saying "the three tools that won us the war in Europe were the Dakota and the landing craft and the Jeep". In 1940, the American Defense Department sent out a tough spec for a military workhorse. Many companies took one look at the seemingly impossible specification and 49-day deadline and turned it down flat. The design that won the tender and made it into production and the history books was a mixture of the ideas and abilities of Ford, Bantam, and Willys-Overland. A stunning triumph of function over form, the Jeep not only won the war, but went on to become a cult off-roader that's still with us now. The Willys Jeep is surely the most original 4x4 by far.

POWER
*The hardy
L-head motor
developed 60 bhp.*

SAFETY STRAPS
*Doors would have added weight,
so side straps were a token
gesture towards driver safety.*

TRICKY DRIVE
*High clutch, narrow footwell, and
unmovable seat forced a knees-
splayed driving position.*

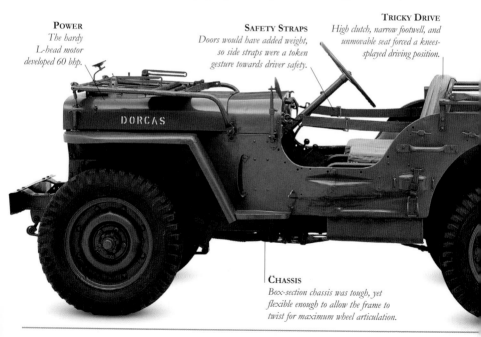

CHASSIS
*Box-section chassis was tough, yet
flexible enough to allow the frame to
twist for maximum wheel articulation.*

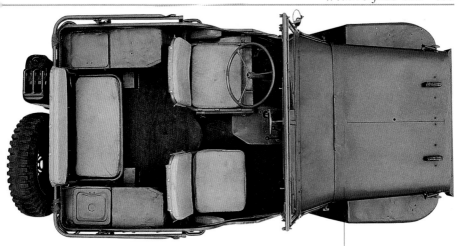

SUSPENSION
Leaf springs and hydraulic shocks gave a surprisingly good ride.

EXPOSED COLUMN
Driver safety wasn't a Jeep strong point. Many GIs ended up impaled on the steering column even after low-speed impacts.

THIRSTY
The Jeep may have had competence, but it also had a prodigious thirst for fuel.

LIFESPAN
The Jeep was a brilliantly simple solution to the problem of mobility at war, but the life expectancy of an average vehicle was expected to be less than a week!

JEEP NAME
Jeeps were first called General Purpose cars, then MA, and finally MB, but nobody's sure of the origins of the Jeep name. Some say it is a corruption of GP, or General Purpose, others that it was named after Eugene the Jeep, a character in a 1936 Popeye cartoon.

SPECIFICATIONS

MODEL Willys Jeep MB (1943)

PRODUCTION 586,000 (during World War II)

BODY STYLE Open utility vehicle.

CONSTRUCTION Steel body and chassis.

ENGINE 134cid straight-four.

POWER OUTPUT 60 bhp.

TRANSMISSION Three-speed manual, four-wheel drive.

SUSPENSION Leaf springs front and rear.

BRAKES Front and rear drums.

MAXIMUM SPEED 105 km/h (65 mph)

0–60 MPH (0–96 KM/H) 22 sec

A.F.C. 5.7 km/l (16 mpg)

ENGINE
Power was from a Ford straight-four, which took the Jeep to around 105 km/h (65 mph), actually exceeding US Army driving regulations.

RAD CHANGES
Earlier Jeeps had a slatted radiator grille instead of the later pressed-steel bars, as here. The silhouette was low, but ground clearance high to allow driving in streams as deep as 53 cm (21 in). Weather protection was vestigial.

CLUTCH
Quick-release clutch disengaged engine fan for fording streams and rivers.

GEARBOX
The Warner three-speed manual box was supplemented by controls allowing the driver to select two- or four-wheel drive in high or low ratios.

FRONT VIEW
The Jeep's bonnet was secured using quick-release sprung catches. The upper catch held the fold-down windscreen. Those stark wings and large all-terrain tyres may look humble and functional, but the Jeep's claim to fame is that it spawned utility vehicles from Nissans and Isuzus to Discoverys and Range Rovers.

WIPERS
Hand-operated windscreen wipers.

GEAR LEVER
First production Jeep model, the MA, had a column change.

EXTRAS
Jeeps came with petrol can, shovel, and long-handled axe.

SPARTAN INTERIOR
Only the generals fought the war in comfort, and Jeep accommodation was strictly no frills. Very early Jeeps have no glove compartment.

HEADLIGHT
The dual-purpose headlight could be rotated back to illuminate the engine bay, which was very useful during night-time manoeuvres.

JOINT EFFORT
Willys and Ford Jeeps saw service in every theatre of war, and the two versions were almost identical. By August 1945, when wartime production of the Jeep ended, the two companies together had manufactured over 600,000 Jeeps. The US Army still carried on using Jeeps well into the Sixties.

INDEX

A

470 INDEX